Power Quality in Electrical Systems

Power Quality in Electrical Systems

Alexander Kusko, Sc.D., P.E.

Marc T. Thompson, Ph.D.

New York   Chicago   San Francisco   Lisbon   London   Madrid
Mexico City   Milan   New Delhi   San Juan   Seoul
Singapore   Sydney   Toronto

The McGraw-Hill Companies

Library of Congress Cataloging-in-Publication Data

Kusko, Alexander, date.
  Power quality in electrical systems / Alexander Kusko, Marc T.
Thompson.
      p.   cm.
  Includes bibliographical references and index.
  ISBN-13: 978-0-07-147075-9 (alk. paper)
  ISBN-10: 0-07-147075-1 (alk. paper)
  1. Electric power systems—Electric losses.   2. Electric power
distribution.   I. Thompson, Marc T.   II. Title.
  TK1005.K874   2007
  621.319—dc22   2007016283

1 2 3 4 5 6 7 8 9 0   DOC/DOC   0 1 3 2 1 0 9 8 7

ISBN-13: 978-0-07-147075-9
ISBN-10: 0-07-147075-1

Sponsoring Editor: Stephen S. Chapman
Production Supervisor: Pamela A. Pelton
Editing Supervisor: Stephen M. Smith
Project Manager: Rasika Mathur
Copy Editor: Mike McGee
Proofreader: Ragini Pandey
Indexer: WordCo Indexing Services
Art Director, Cover: Jeff Weeks
Composition: International Typesetting and Composition

Printed and bound by RR Donnelley.

McGraw-Hill books are available at special quantity discounts to use as premiums
and sales promotions, or for use in corporate training programs. For more information, please write to the Director of Special Sales, McGraw-Hill Professional,
Two Penn Plaza, New York, NY 10121-2298. Or contact your local bookstore.

This book is printed on acid-free paper.

ABOUT THE AUTHORS

ALEXANDER KUSKO, SC.D., P.E., is Corporate Vice President of Exponent. He was formerly an associate professor of engineering at MIT. Dr. Kusko is a Life Fellow of the IEEE and served on the committee for the original IEEE Standard 519-1981 on Harmonic Control in Electrical Power Systems.

MARC T. THOMPSON, PH.D., is President of Thompson Consulting, Inc., an engineering consulting firm specializing in power electronics, magnetic design, and analog circuits and systems. He is also an adjunct professor of electrical engineering at Worcester Polytechnic Institute and a firefighter with the Harvard (Massachusetts) Fire Department.

Contents

Preface

This book is intended for use by practicing power engineers and managers interested in the emerging field of power quality in electrical systems. We take a real-world point of view throughout with numerous examples compiled from the literature and the authors' engineering experiences.

Acknowledgments

PSPICE simulations were done using the Microsim Evaluation version 8.0.

The authors gratefully acknowledge the cooperation of the IEEE with regard to figures reprinted from IEEE standards, and with permission from the IEEE. The IEEE disclaims any responsibility or liability resulting from the placement and use in the described manner.

ALEXANDER KUSKO, SC.D., P.E.

MARC T. THOMPSON, PH.D.

Power Quality in Electrical Systems

1

Introduction

In this introductory chapter, we shall attempt to define the term "power quality," and then discuss several power-quality "events." Power-quality "events" happen during fault conditions, lightning strikes, and other occurrences that adversely affect the line-voltage and/or current waveforms. We shall define these events and their causes, and the possible ramifications of poor power quality.

Background

In recent years, there has been an increased emphasis on, and concern for, the quality of power delivered to factories, commercial establishments, and residences [1.1–1.15]. This is due in part to the preponderance of harmonic-creating systems in use. Adjustable-speed drives, switching power supplies, arc furnaces, electronic fluorescent lamp ballasts, and other harmonic-generating equipment all contribute to the harmonic burden the system must accommodate [1.15–1.17]. In addition, utility switching and fault clearing produce disturbances that affect the quality of delivered power. In addressing this problem, the Institute of Electrical and Electronics Engineers (IEEE) has done significant work on the definition, detection, and mitigation of power-quality events [1.18–1.27].

Much of the equipment in use today is susceptible to damage or service interruption during poor power-quality events [1.28]. Everyone with a computer has experienced a computer shutdown and reboot, with a loss of work resulting. Often, this is caused by poor power quality on the 120-V line. As we'll see later, poor power quality also affects the efficiency and operation of electric devices and other equipment in factories and offices [1.29–1.30].

Various health organizations have also shown an increased interest in stray magnetic and electric fields, resulting in guidelines on the levels of these fields [1.31]. Since currents create magnetic fields, it is possible to lessen AC magnetic fields by reducing harmonic currents present in the line-voltage conductors.

Harmonic pollution on a power line can be quantified by a measure known as *total harmonic distortion* or THD.[1] High harmonic distortion can negatively impact a facility's electric distribution system, and can generate excessive heat in motors, causing early failures. Heat also builds up in wire insulation causing breakdown and failure. Increased operating temperatures can affect other equipment as well, resulting in malfunctions and early failure. In addition, harmonics on the power line can prompt computers to restart and adversely affect other sensitive analog circuits.

The reasons for the increased interest in power quality can be summarized as follows [1.32]:

- **Metering:** Poor power quality can affect the accuracy of utility metering.

- **Protective relays:** Poor power quality can cause protective relays to malfunction.

- **Downtime:** Poor power quality can result in equipment downtime and/or damage, resulting in a loss of productivity.

- **Cost:** Poor power quality can result in increased costs due to the preceding effects.

- **Electromagnetic compatibility:** Poor power quality can result in problems with electromagnetic compatibility and noise [1.33–1.39].

Ideal Voltage Waveform

Ideal power quality for the source of energy to an electrical load is represented by the single-phase waveform of voltage shown in Figure 1.1 and the three-phase waveforms of voltage shown in Figure 1.2. The amplitude, frequency, and any distortion of the waveforms would remain within prescribed limits.

When the voltages shown in Figure 1.1 and Figure 1.2 are applied to electrical loads, the load currents will have frequency and amplitudes dependent on the impedance or other characteristics of the load. If the waveform of the load current is also sinusoidal, the load is termed "linear." If the waveform of the load current is distorted, the load is termed "nonlinear." The load current with distorted waveform can produce

---

[1] THD and other metrics are discussed in Chapter 4.

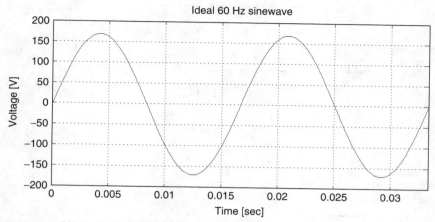

Figure 1.1 Ideal single-phase voltage waveform. The peak value is +170 V, the rms value is 120 V, and the frequency is 60 Hz.

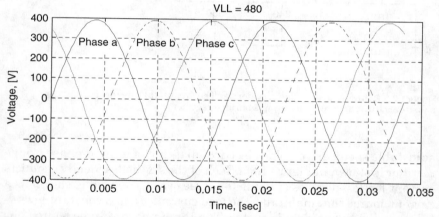

Figure 1.2 An ideal three-phase voltage waveform at 60 Hz with a line-line-voltage of 480 V rms. Shown are the line-neutral voltages of each phase.[2]

distortion of the voltage in the supply system, which is an indication of poor power quality.

Nonlinear Load: The Rectifier

The rectifier, for converting alternating current to direct current, is the most common nonlinear load found in electrical systems. It is used in equipment that ranges from 100-W personal computers to 10,000-kW

---

[2] The line-line-voltage is 480 volts rms; the line-neutral voltage for each phase is $480/\sqrt{3} = 277$ V. Therefore, the peak value for each line-neutral voltage is $277 \text{ V} \times \sqrt{2} = 392$ V.

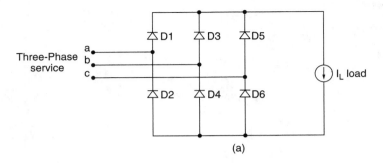

(a)

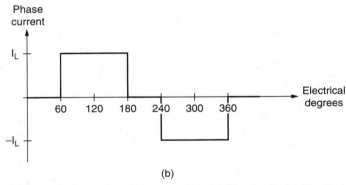

(b)

Figure 1.3   A three-phase bridge rectifier. (a) The circuit. (b) The ideal phase current drawn by a three-phase bridge rectifier.

adjustable speed drives. The electrical diagram of a three-phase bridge rectifier is shown in Figure 1.3a. Each of the six diodes ideally conducts current for 120 degrees of the 360-degree cycle. The load is shown as a current source that maintains the load current, I_L, at a constant level—for example, by an ideal inductor. The three-phase voltage source has the waveform of Figure 1.2. The resultant current in one source phase is shown in Figure 1.3b. The current is highly distorted, as compared to a sine wave, and can distort the voltages of the supply system.

As will be discussed in Chapter 4, the square-wave rectifier load current is described by the Fourier series as a set of harmonic currents. In the case of a three-phase rectifier,[3] the components are the fundamental, and the 5th, 7th, 11th, 13th (and so on) harmonics. The triplens[4] are eliminated. Each of the harmonic currents is treated independently in power-quality analysis.

---

[3] Often called a "six pulse" rectifier.

[4] Triplen (or "triple-n") are harmonics with numbers 3, 9, and so on.

IEEE[5] Standard 519 (IEEE Std. 519-1992) was introduced in 1981 (and updated in 1992) and offers recommended practices for controlling harmonics in electrical systems [1.21]. The IEEE has also released IEEE Standard 1159 (IEEE Std. 1159-1995), which covers recommended methods for measuring and monitoring power quality [1.23].

As time goes on, more and more equipment is being used that creates harmonics in power systems. Conversely, more and more equipment is being used that is susceptible to malfunction due to harmonics. Computers, communications equipment, and other power systems are all susceptible to malfunction or loss of efficiency due to the effects of harmonics.

For instance, in electric motors, harmonic current causes AC losses in the core and copper windings.[6] This can result in core heating, winding heating, torque pulsations, and loss of efficiency in these motors. Harmonics can also result in an increase in audible noise from motors and transformers[7] and can excite mechanical resonances in electric motors and their loads.

Harmonic voltages and currents can also cause false tripping of ground fault circuit interrupters (GFCIs). These devices are used extensively in residences for local protection near appliances. False triggering of GFCIs is a nuisance to the end user.

Instrument and relay transformer accuracy can be affected by harmonics, which can also cause nuisance tripping of circuit breakers. Harmonics can affect metering as well, and may prompt both negative and positive errors.

High-frequency switching circuits—such as switching power supplies, power factor correction circuits, and adjustable-speed drives—create high-frequency components that are not at multiples of line frequency. For instance, a switching power supply operating at 75 kHz produces high-frequency components at integer multiples of the fundamental 75 kHz switching frequency, as shown in Figure 1.4. These frequency components are sometimes termed "interharmonics" to differentiate them from harmonics, which are multiples of the line frequency. Other worldwide standards specify the amount of harmonic noise that can be injected into a power line. IEC-1000-2-1 [1.40] defines interharmonics as follows:

> Between the harmonics of the power frequency voltage and current, further frequencies can be observed which are not an integer of the fundamental. They can appear as discrete frequencies or as a wide-band spectrum.

---

[5] Institute of Electrical and Electronics Engineers.

[6] The losses in the copper winding are due to skin-effect phenomena. Losses in the core are due to eddy currents as well as "hysteresis" loss.

[7] IEEE Std. C57.12.00-1987 recommends a current distortion factor of less than 5 percent for transformers.

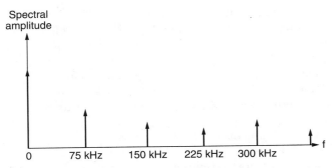

Figure 1.4 Typical interharmonic spectra produced by a high-frequency switching power supply with switching frequency 75 kHz. We see interharmonics at multiples of 75 kHz.

Other sources of interharmonics are cycloconverters, arc furnaces, and other loads that do not operate synchronously with the power-line frequency [1.41].

High-frequency components can interfere with other electronic systems nearby and also contribute to radiated electromagnetic interference (EMI). Medical electronics is particularly susceptible to the effects of EMI due to the low-level signals involved. Telephone transmission can be disrupted by EMI-induced noise.

This recent emphasis on the purity of delivered power has resulted in a new field of study—that of "power quality."

The Definition of Power Quality

Power quality, loosely defined, is the study of powering and grounding electronic systems so as to maintain the integrity of the power supplied to the system. IEEE Standard 1159[8] defines power quality as [1.23]:

> The concept of powering and grounding sensitive equipment in a manner that is suitable for the operation of that equipment.

Power quality is defined in the IEEE 100 Authoritative Dictionary of IEEE Standard Terms as ([1.42], p. 855):

> The concept of powering and grounding electronic equipment in a manner that is suitable to the operation of that equipment and compatible with the premise wiring system and other connected equipment.

---

[8] IEEE Std. 1159-1995, section 3.1.47, p. 5.

Equally authoritative, the qualification is made in the *Standard Handbook of Electrical Engineers*, 14th edition, (2000) ([1.43] pp. 18–117):

> Good power quality, however, is not easy to define because what is good power quality to a refrigerator motor may not be good enough for today's personal computers and other sensitive loads. For example, a short (momentary) outage would not noticeably affect motors, lights, etc. but could cause a major nuisance to digital clocks, videocassette recorders (VCRs) etc.

Examples of poor power quality

Poor power quality is usually identified in the "powering" part of the definition, namely in the deviations in the voltage waveform from the ideal of Figure 1.1. A set of waveforms for typical power disturbances is shown in Figure 1.5. These waveforms are either (a) observed, (b) calculated, or (c) generated by test equipment.

The following are some examples of poor power quality and descriptions of poor power-quality "events." Throughout, we shall paraphrase the IEEE definitions.

- A voltage sag (also called a "dip"[9]) is a brief decrease in the rms line-voltage of 10 to 90 percent of the nominal line-voltage. The duration of a sag is 0.5 cycle to 1 minute [1.44–1.50]. Common sources of sags are the starting of large induction motors and utility faults.

- A voltage swell is the converse to the sag. A swell is a brief increase in the rms line-voltage of 110 to 180 percent of the nominal line-voltage for a duration of 0.5 cycle to 1 minute. Sources of voltage swells are line faults and incorrect tap settings in tap changers in substations.

- An impulsive transient is a brief, unidirectional variation in voltage, current, or both on a power line. The most common causes of impulsive transients are lightning strikes, switching of inductive loads, or switching in the power distribution system. These transients can result in equipment shutdown or damage if the disturbance level is high enough. The effects of transients can be mitigated by the use of transient voltage suppressors such as Zener diodes and MOVs (metal-oxide varistors).

- An oscillatory transient is a brief, bidirectional variation in voltage, current, or both on a power line. These can occur due to the switching of power factor correction capacitors, or transformer ferroresonance.

- An interruption is defined as a reduction in line-voltage or current to less than 10 percent of the nominal, not exceeding 60 seconds in length.

---

[9] Generally, it's called a sag in the U.S. and a dip in the UK.

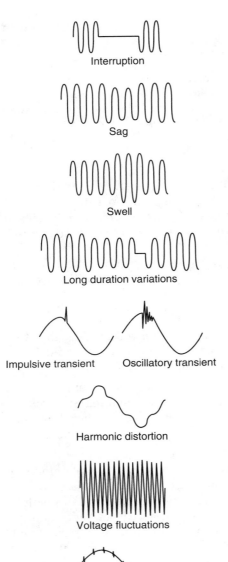

Interruption

Sag

Swell

Long duration variations

Impulsive transient   Oscillatory transient

Harmonic distortion

Voltage fluctuations

Figure 1.5 Typical power disturbances, from [1.2].
[© 1997 IEEE, reprinted with
permission]

Noise

- Another common power-quality event is "notching," which can be created by rectifiers that have finite line inductance. The notches show
  up due to an effect known as "current commutation."

- Voltage fluctuations are relatively small (less than ±5 percent) variations in the rms line-voltage. These variations can be caused by

cycloconverters, arc furnaces, and other systems that draw current not in synchronization with the line frequency [1.51–1.61]. Such fluctuations can result in variations in the lighting intensity due to an effect known as "flicker" which is visible to the end user.

- A voltage "imbalance" is a variation in the amplitudes of three-phase voltages, relative to one another.

The need for corrections

Why do we need to detect and/or correct power-quality events [1.63–1.64]? The bottom line is that the end user wants to see the non-interruption of good quality electrical service because the cost of downtime is high. Shown in Table 1.1, we see a listing of possible mitigating strategies for poor power quality, and the relative costs of each.

The Scope of This Text

We will address the significant aspects of power quality in the following chapters:

Chapter 1, *Introduction*, provides a background for the subject, including definitions, examples, and an outline for the book.

Chapter 2, *Power Quality Standards*, discusses various power-quality standards, such as those from the IEEE and other bodies. Included are standards discussing harmonic distortion (frequencies that are multiples of the line frequency) as well as high-frequency interharmonics caused by switching power supplies, inverters, and other high-frequency circuits.

Chapter 3, *Voltage Distortion*, discusses line-voltage distortion, and its causes and effects.

Chapter 4, *Harmonics*, is an overall discussion of the manner in which line-voltage and line-current distortion are described in quantitative terms using the concept of harmonics and the Fourier series, and spectra of periodic waveforms.

Chapter 5, *Harmonic Current Sources*, discusses sources of harmonic currents. This equipment, such as electronic converters, creates frequency components at multiples of the line frequency that, in turn, cause voltage distortion.

Chapter 6, *Power Harmonic Filters*, discusses power harmonic filters, a class of equipment used to reduce the effect of harmonic currents and improve the quality of the power provided to loads. These filters can be either passive or active.

TABLE 1.1 The Relative Cost of Mitigating Voltage Disturbances [1.2] [© 1997 IEEE, reprinted with permission]

Mitigating device	Application of device	Relative cost of device
Solid State Transfer Switch	Device used in conjunction with an alternate electrical supply. Depending on the speed and quality of the transfer, this switch can be used in cases of interruptions, sags, swells, and long duration overvoltages and undervoltages. (Medium and high voltage)	$300/kVA
Standby Generator	A Standby Generator is used to supply the electrical power as an alternate to the normal power supplied by the utility. A means of transition is required. Generators are used in cases of interruptions.	$260 – $500/kVA
Uninterruptible Power Supply (UPS)	UPS equipment can be used in cases of interruptions, sags, swells, and voltage fluctuations. Some success can also be achieved in instances of impulsive and oscillatory transients, long duration overvoltages and undervoltages, and noise.	$1,000 – $3,000/kVA
Superconducting Storage Device (SSD)	Device utilizes energy storage within a magnet that is created by the flow of DC current. Utilized for interruptions, and sags.	$1,000/kVA
Motor-Generator (MG) Sets	Equipment used in all cases except long duration outages. Motor drives generator which isolates output power from incoming source.	$600/kVA
Reduced Voltage Starters	These devices are used to reduce the current inrush at motor start-up and thus lessen the voltage sag associated with that current inrush.	$25 – $50/kVA
Contactor Ride-through Devices	Developing technology aimed at holding a constant voltage across contactor coils and thus ride through a voltage sag.	$30/contactor
Ferroresonant (CVT) Transformers	These devices utilize ferroresonant technology and transformer saturation for success in cases of sags, swells, and long duration undervoltages.	$400 – $1,500/kVA
Surge Protective Devices (SPD)	Device used to address impulsive transients. Some success with oscillatory transients.	$50 – $200/kVA
Shielded Isolation Transformers	These devices are effective in cases of oscillatory transients and noise. Some success in cases of impulsive transients.	$20 – $60/kVA
Line Reactors	These devices are effective in cases of oscillatory transients.	$15 – $100/kVA
K-Factor Transformers	Device used where harmonics may be present.	$60 – $100/kVA
Harmonic Filters	Device used to provide a low impedance path for harmonic currents. Reactors used in conjunction with (power factor correction) capacitors.	$75 – $250/kVA
Fiberoptes Cable	Alternative to copper cabling where communications may be susceptible to noise.	N/A
Optical Isolators	Supplement to copper cabling where communications may be susceptible to noise.	N/A
Noise Filters	Device used to pass 60 Hz power signal and block unwanted (noise) frequencies.	N/A

Chapter 7, *Switch Mode Power Supplies*, discusses switching power supplies that are incorporated in every personal computer, server, industrial controllers, and other electronic equipment, and which create high-frequency components that result in electromagnetic interference (EMI).

Chapter 8, *Methods for Correction of Power-Quality Problems*, is a preliminary look at methods for design of equipment and supply systems to correct for effects of poor power quality.

Chapter 9, *Uninterruptible Power Supplies*, discusses the most widely employed equipment to prevent poor power quality of the supply system from affecting sensitive loads.

Chapter 10, *Dynamic Voltage Compensators*, is a description of low-cost equipment to prevent the most frequent short-time line-voltage dips from affecting sensitive equipment.

Chapter 11, *Power-Quality Events*, discusses how power-quality events, such as voltage sags and interruptions affect personal computers and other equipment.

Chapter 12, *Adjustable Speed Drives (ASDs) and Induction Motors*, discusses major three-phase power-electronic equipment that both affect power quality and are affected *by* poor power quality.

Chapter 13, *Standby Power Systems*, consisting of UPSs, discusses engine-generator and transfer switches to supply uninterrupted power to critical loads such as computer data centers.

Chapter 14, *Measurements*, discusses methods and equipment for performing power-quality measurements.

Comment on References

The business of electrical engineering is to, first, provide "clean" uninterrupted electric power to all customers and, second, to design and manufacture equipment that will operate with the actual power delivered. As such, practically all of the electrical engineering literature bears on power quality. A group of pertinent references is given at the end of this chapter and in the following chapters of the book. Two important early references that defined the field are the following:

- "IEEE Recommended Practice for Emergency and Standby Power Systems for Industrial and Commercial Applications," (The Orange Book), IEEE Std. 446-1995 [1.20]

- "IEEE Recommended Practices and Requirements for Harmonic Control in Electrical Power Systems," IEEE Std. 519-1992, revision of IEEE Std. 519-1981 [1.21]

References

[1.1] J. M. Clemmensen and R. J. Ferraro, "The Emerging Problem of Electric Power Quality," *Public Utilities Fortnightly*, November 28, 1985.

[1.2] J. G. Dougherty and W. L. Stebbins, "Power Quality: A Utility and Industry Perspective," *Proceedings of the IEEE 1997 Annual Textile, Fiber and Film Industry Technical Conference*, May 6–8, 1997, pp. 1–10.

[1.3] Dranetz-BMI, *The Dranetz-BMI Field Handbook for Power Quality Analysis*, Dranetz-BMI, 1998.

[1.4] R. C. Dugan, M. F. McGranaghan, S. Santoso, and H. W. Beaty, *Electrical Power Systems Quality*, McGraw-Hill, 2003.

[1.5] R. A. Flores, "State of the Art in the Classification of Power Quality Events, an Overview," *Proceedings of the 2002 10th International Conference on Harmonics and Quality of Power*, pp. 17–20.

[1.6] G. T. Heydt, "Electric Power Quality: A Tutorial Introduction," *IEEE Computer Applications in Power*, vol. 11, no. 1, January 1998, pp. 15–19.

[1.7] M. A. Golkar, "Electric Power Quality: Types and Measurements," *2004 IEEE International Conference on Electric Utility Deregulation, Restructuring and Power Technologies (DRPT2004)*, April 2004, Hong Kong, pp. 317–321.

[1.8] T. E. Grebe, "Power Quality and the Utility/Customer Interface," *SOUTHCON '94 Conference Record*, March 29–31, 1994, pp. 372–377.

[1.9] T. Ise, Y. Hayashi, and K. Tsuji, "Definitions of Power Quality Levels and the Simplest Approach for Unbundled Power Quality Services," *Proceedings of the Ninth International Conference on Harmonics and Quality of Power*, October 1–4, 2000, pp. 385–390.

[1.10] B. Kennedy, *Power Quality Primer*, McGraw-Hill, 2000.

[1.11] S. D. MacGregor, "An Overview of Power Quality Issues and Solutions," *Proceedings of the 1998 IEEE Cement Industry Technical Conference*, May 17–21, 1998, pp. 57–64.

[1.12] F. D. Martzloff and T. M. Gruzs, "Power Quality Site Surveys: Facts, Fiction and Fallacies," *IEEE Transactions on Industry Applications*, vol. 24, no. 6, November/December 1988.

[1.13] J. Seymour and T. Horsley, "The Seven Types of Power Problems," APC Whitepaper #18. Available on the Web at http://www.apcmedia.com/salestools/VAVR-5WKLPK_R0_EN.pdf.

[1.14] J. Stones and A. Collinson, "Power Quality," *Power Engineering Journal*, April, 2001, pp. 58–64

[1.15] IEEE, "Interharmonics in Power Systems," IEEE Interharmonic Task Force. Available on the Web at http://grouper.ieee.org/groups/harmonic/iharm/docs/ihfinal.pdf.

[1.16] Y. Yacamini, "Power Systems Harmonics—Part 1: Harmonic Sources," *Power Engineering Journal*, August 1994, pp. 193–198.

[1.17] ____, "Power Systems Harmonics—Part 3: Problems Caused by Distorted Supplies," *Power Engineering Journal*, October 1995, pp. 233–238.

[1.18] C. K. Duffey and R. P. Stratford "Update of Harmonic Standard IEEE-519: IEEE Recommended Practices and Requirements for Harmonic Control in Electric Power Systems," *IEEE Transactions on Industry Applications*, vol. 25, no. 6, November/December 1989, pp. 1025–1034.

[1.19] T. Hoevenaars, K. LeDoux, and M. Colosino, "Interpreting IEEE Std. 519 and Meeting its Harmonic Limits in VFD Applications," *Proceedings of the IEEE Industry Applications Society 50th Annual Petroleum and Chemical Industry Conference*, September 15–17, 2003, pp. 145–150.

[1.20] IEEE, "IEEE Recommended Practice for Emergency and Standby Power Systems for Industrial and Commercial Applications," (The Orange Book), IEEE Std. 446-1995

[1.21] ____, "IEEE Recommended Practices and Requirements for Harmonic Control in Electrical Power Systems," IEEE Std. 519-1992, revision of IEEE Std. 519-1981.

[1.22] ____, "IEEE Recommended Practice for Powering and Grounding Sensitive Electronic Equipment," IEEE Std. 1100-1992 (Emerald Book).

[1.23] ____, "IEEE Recommended Practice for Monitoring Electric Power Quality," IEEE Std. 1159-1995.

[1.24] ____, "IEEE Guide for Service to Equipment Sensitive to Momentary Voltage Disturbances," IEEE Std. 1250-1995.

[1.25] ____, "IEEE Recommended Practice for Evaluating Electric Power System Compatibility with Electronic Process Equipment," IEEE Std. 1346-1998.

[1.26] IEEE, "IEEE Recommended Practice for Measurement and Limits of Voltage Fluctuations and Associated Light Flicker on AC Power Systems," IEEE Std. 1453-2004.

[1.27] M. E. Baran, J. Maclaga, A. W. Kelley, and K. Craven, "Effects of Power Disturbances on Computer Systems," *IEEE Transactions on Power Delivery*, vol. 13, no. 4, October 1998, pp. 1309–1315.

[1.28] K. Johnson and R. Zavadil, "Assessing the Impacts of Nonlinear Loads on Power Quality in Commercial Buildings—An Overview," *Conference Record of the 1991 IEEE Industry Applications Society Annual Meeting*, September 28–October 4, 1991, pp. 1863–1869.

[1.29] V. E. Wagner, "Effects of Harmonics on Equipment," *IEEE Transactions on Power Delivery*, vol. 8, no. 2, April 1993, pp. 672–680.

[1.30] Siemens, "Harmonic Distortion Damages Equipment and Creates a Host of Other Problems." Whitepaper available on the Web at http://www.sbt.siemens.com/HVP/Components/Documentation/SI033WhitePaper.pdf.

[1.31] ICNIRP, "Guidelines for Limiting Exposure to Time-Varying Electric, Magnetic and Electromagnetic Fields (Up to 300 MHz)," International Commission on Non-Ionizing Radiation Protection.

[1.32] R. Redl and A. S. Kislovski, "Telecom Power Supplies and Power Quality," *Proceedings of the 17th International Telecommunications Energy Conference, INTELEC '95*, October 29–November 1, 1995, pp. 13–21.

[1.33] K. Armstrong, "Filters," *Conformity*, 2004, pp. 126–133.

[1.34] ____, "Spotlight on Filters," *Conformity*, July 2003, pp. 28–32.

[1.35] ANSI, "American National Standard Guide on the Application and Evaluation of EMI Power-Line Filters for Commercial Use," *ANSI C63.13 1991*.

[1.36] Astec, Inc., "EMI Suppression," Application note 1821, November 12, 1998.

[1.37] J. R. Barnes, "Designing Electronic Systems for ESD Immunity," *Conformity*, February 2003, pp. 18–27.

[1.38] H. Chung, S. Y. R. Hui, and K. K. Tse, "Reduction of Power Converter EMI Emission Using Soft-Switching Technique," *IEEE Transactions on Electromagnetic Compatibility*, vol. 40, no. 3, August 1998, pp. 282–287.

[1.39] T. Curatolo and S. Cogger, "Enhancing a Power Supply to Ensure EMI Compliance," *EDN*, February 17, 2005, pp. 67–74.

[1.40] CEI/IEC 1000-2-1: 1990, "Electromagnetic Compatibility," 1st ed, 1990.

[1.41] IEEE, "Interharmonics in Power Systems," *IEEE Interharmonics Task Force*, December 1, 1997.

[1.42] ____, *IEEE 100 The Authoritative Dictionary of IEEE Standard Terms*, Standards Information Network, IEEE Press.

[1.43] D. G. Fink and H. W. Beaty, eds., *Standard Handbook for Electrical Engineers*, McGraw-Hill, 1999.

[1.44] M. F. Alves and T. N. Ribeiro, "Voltage Sag: An Overview of IEC and IEEE Standards and Application Criteria," *Proceedings of the 1999 IEEE Transmission and Distribution Conference*, April 11–16, 1999, pp. 585–589.

[1.45] M. Bollen, "Voltage Sags: Effects, Prediction and Mitigation," *Power Engineering Journal*, June 1996, pp. 129–135.

[1.46] M. S. Daniel, "A 35-kV System Voltage Sag Improvement," *IEEE Transactions on Power Delivery*, vol. 19, no. 1, January 2004, pp. 261–265.

[1.47] S. Djokic, J. Desmet, G. Vanalme, J. V. Milanovic, and K. Stockman, "Sensitivity of Personal Computers to Voltage Sags and Short Interruptions," *IEEE Transactions on Power Delivery*, vol. 20, no. 1, January 2005, pp. 375–383.

[1.48] K. J. Kornick, and H. Q. Li, "Power Quality and Voltage Dips: Problems, Requirements, Responsibilities," *Proceedings of the 5th International Conference on Advances in Power System Control, Operation and Management, APSCOM 2000*, Hong Kong, October 2000, pp. 149–156.

[1.49] J. Lamoree, D. Mueller, P. Vinett, W. Jones, and M. Samotyj, "Voltage Sag Analysis Case Studies," *IEEE Transactions on Industry Applications*, vol. 30, no. 4, July/August 1994, pp. 1083–1089.

[1.50] PG&E, "Short Duration Voltage Sags Can Cause Disturbances." Available on the Web at http://www.pge.com/docs/pdfs/biz/power_quality/power_quality_notes/voltagesags.pdf.

[1.51] B. Bhargava, "Arc Furnace Flicker Measurements and Control," IEEE Transactions on Power Delivery, vol. 8, no. 1, January 1993, pp. 400–410.

[1.52] G. C. Cornfield, "Definition and Measurement of Voltage Flicker," *Proceedings of the IEE Colloquium on Electronics in Power Systems Measurement*, April 18, 1988, pp. 4/1–4/4.

[1.53] M. De Koster, E. De Jaiger, and W. Vancoistem, "Light Flicker Caused by Interharmonic." Available on the Web at http://grouper.ieee.org/groups/harmonic/iharm/docs,ihflicker.pdf.

[1.54] A. E. Emanuel, and L. Peretto, "A Simple Lamp-Eye-Brain Model for Flicker Observations," *IEEE Transactions on Power Delivery*, vol. 19, no. 3, July 2004, pp. 1308–1313.

[1.55] D. Gallo, C. Landi, and N. Pasquino, "An Instrument for the Objective Measurement of Light Flicker," *IMTC 2005—Instrumentation and Measurement Technology Conference*, Ottawa, Canada, May 17–19, 2005, pp. 1942–1947.

[1.56] D. Gallo, R. Langella, and A. Testa, "Light Flicker Prediction Based on Voltage Spectral Analysis," Proceedings of the 2001 IEEE Porto Power Tech Conference, September 10–13, 2001, Porto, Portugal.

[1.57] D. Gallo, C. Landi, R. Langella, and A. Testa, "IEC Flickermeter Response to Interharmonic Pollution," *2004 11th International Conference on Harmonics and Quality of Power*, September 12–15, 2004, pp. 489–494.

[1.58] A. A. Girgis, J. W. Stephens, and E. B. Makram, "Measurement and Prediction of Voltage Flicker Magnitude and Frequency," *IEEE Transactions on Power Delivery*, vol. 10, no. 3, July 1995, pp. 1600–1605.

[1.59] I. Langmuir, "The Flicker of Incandescent Lamps on Alternating Current Circuits and Stroboscopic Effects," *GE Review*, vol. 17, no. 3, March 1914, pp. 294–300.

[1.60] E. L. Owen, "Power Disturbance and Quality: Light Flicker Voltage Requirements," IEEE Industry Applications Magazine, vol. 2, no. 1, January–February 1996, pp. 20–27.

[1.61] C.-S. Wang and M. J. Devaney, "Incandescent Lamp Flicker Mitigation and Measurement," *IEEE Transactions on Instrumentation and Measurement*, vol. 53, no. 4, August 2004, pp. 1028–1034.

[1.62] K. N. Sakthivel, S. K. Das, and K. R. Kini, "Importance of Quality AC Power Distribution and Understanding of EMC Standards IEC 61000-3-2, IEC 61000–3–3, and IEC 61000-3-11," *Proceedings of the 8th International Conference on Electromagnetic Interference and Compatibility, INCEMIC 2003*, December 18–19, 2003, pp. 423–430.

[1.63] F. J. Salem and R. A. Simmons, "Power Quality from a Utility Perspective," *Proceedings of the Ninth International Conference on Harmonics and Quality of Power*, October 1–4, 2000, pp. 882–886.

[1.64] R. C. Sermon, "An Overview of Power Quality Standards and Guidelines from the End-User's Point-of-View," *Proceedings of the 2005 Rural Electric Power Conference*, May 8–10, 2005, pp. B1-1–B1-5.

Power-Quality Standards

This chapter offers some details on various standards addressing the issues of power quality in electric systems. Standards are needed so all end users (industrial, commercial, and residential) and transmission and distribution suppliers (the utilities) speak the same language when discussing power-quality issues. Standards also define recommended limits for events that degrade power quality.

IEEE Standards 519 and 1159

IEEE Standards are publications that provide acceptable design practice. IEEE Standards addressing power quality include those defining acceptable power quality (IEEE Standard 519) and another standard relating to the measurement of power-quality "events" (IEEE Standard 1159). In later chapters of this book, we'll use several figures from the IEEE Standards so the reader will have a flavor for the coverage. Both of these standards focus on AC systems and their harmonics (that is, multiples of the line frequency).

IEEE Standard 519 [2.1] (denoted IEEE Std. 519-1992) is titled "IEEE Recommended Practices and Requirements for Harmonic Control in Electrical Power Systems." The abstract of this standard notes that power conversion units are being used today in industrial and commercial facilities, and there are challenges associated with harmonics and reactive power control of such systems. The standard covers limits to the various disturbances recommended to the power distribution system. The 1992 standard is a revision of an earlier IEEE work published in 1981 covering harmonic control.

The basic themes of IEEE Standard 519 are twofold. First, the utility has the responsibility to produce good quality voltage sine waves.

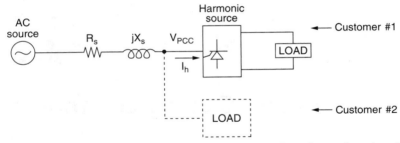

Figure 2.1 Harmonic-generating load causing voltage distortion at the point of common coupling (PCC). The AC source is modeled as an ideal voltage source in series with a resistance R_s and a reactance jX_s.

Secondly, end-use customers have the responsibility to limit the harmonic currents their circuits draw from the line.

Shown in Figure 2.1 is a utility system feeder serving two customers. The utility source has resistance R and line reactance jX_s. The resistance and reactance model the impedances of the utility source, any transformers and switchgear, and power cabling. Customer #1 on the line draws harmonic current I_h, as shown, perhaps by operating adjustable-speed drives, arc furnaces, or other harmonic-creating systems.

The voltage Customer #2 sees at the service entrance is the voltage at the "point of common coupling," often abbreviated as "PCC." The harmonics drawn by Customer #1 cause voltage distortion at the PCC, due to the voltage drop in the line resistance and reactance due to the harmonic current.

Shown in Figure 2.2 are harmonic distortion limits found in IEEE 519 for harmonic distortion limits at the point of common coupling. The voltage harmonic distortion limits apply to the quality of the power the utility must deliver to the customer. For instance, for systems of less than 69 kV, IEEE 519 requires limits of 3 percent harmonic distortion for an individual frequency component and 5 percent for total harmonic distortion.

Table 11.1
Voltage Distortion Limits

Bus Voltage at PCC	Individual Voltage Distortion (%)	Total Voltage Distortion THD (%)
69 kV and below	3.0	5.0
69.001 kV through 161 kV	1.5	2.5
161.001 kV and above	1.0	1.5

NOTE: High-voltage systems can have up to 2.0% THD where the cause is an HVDC terminal that will attenuate by the time it is tapped for a user.

Figure 2.2 Voltage harmonic distortion limits from IEEE Std. 519-1992, p. 85 [2.1]. [© 1992, IEEE, reprinted with permission]

**Table 10.3
Current Distortion Limits for General Distribution Systems
(120 V Through 69 000 V)**

Maximum Harmonic Current Distortion
in Percent of I_L

Individual Harmonic Order (Odd Harmonics)

I_{SC}/I_L	<11	$11 \leq h < 17$	$17 \leq h < 23$	$23 \leq h < 35$	$35 \leq h$	TDD
<20*	4.0	2.0	1.5	0.6	0.3	5.0
20<50	7.0	3.5	2.5	1.0	0.5	8.0
50<100	10.0	4.5	4.0	1.5	0.7	12.0
100<1000	12.0	5.5	5.0	2.0	1.0	15.0
>1000	15.0	7.0	6.0	2.5	1.4	20.0

Even harmonics are limited to 25% of the odd harmonic limits above.

Current distortions that result in a dc offset, e.g., half-wave converters, are not allowed.

*All power generation equipment is limited to these values of current distortion, regardless of actual I_{SC}/I_L.

Where
  I_{SC} = maximum short-circuit current at PCC.
  I_L = maximum demand load-current (fundamental frequency component) at PCC.

Figure 2.3 Current harmonic distortion limits [2.1].
[© 1992, IEEE, reprinted with permission]

Shown in Figure 2.3 are harmonic distortion limits found in IEEE 519 for current drawn by loads at the point of common coupling. The current harmonic distortion limits apply to limits of harmonics that loads should draw from the utility at the PCC. Note that the harmonic limits differ based on the I_{SC}/I_L rating, where I_{SC} is the maximum short-circuit current at the PCC, and I_L is the maximum demand load current at the PCC.

IEEE Standard 1159 [2.2] is entitled "IEEE Recommended Practice for Monitoring Electric Power Quality," and as its title suggests, this standard covers recommended methods of measuring power-quality events. Many different types of power-quality measurement devices exist and it is important for workers in different areas of power distribution, transmission, and processing to use the same language and measurement techniques. In future chapters, we draw extensively from IEEE Standards 519 and 1159.

ANSI Standard C84

The American National Standards Institute sets guidelines for 120-V service in ANSI Standard C84-1 (1999) [2.3]. Shown in Figure 2.4 are the ranges labeled "A" and "B." Range A is the optimal voltage range, and is 5 percent of nominal voltage. For 120-V service, range A is 114 V

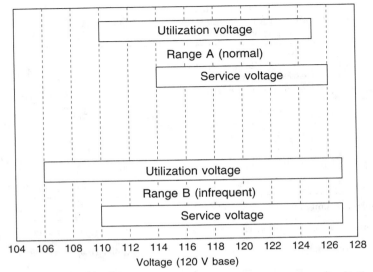

Figure 2.4   Graphical view of ANSI voltage ranges for 120-V service [2.4].
[© 1995, IEEE, reprinted with permission]

to 126 V. Range "B" is acceptable but not optimal, and is in the range of 91.7 percent to 105.8 percent of nominal. This range is allowable for infrequent use. Note that voltage sags and surges go beyond these limits.

CBEMA and ITIC Curves

Computer equipment sensitivity to sags and swells can be charted in curves of acceptable sag/swell amplitude versus event duration. In the 1970s, the Computer Business Equipment Manufacturers Association (CBEMA) developed the curve [2.5] of Figure 2.5 utilizing historical data from mainframe computer operations, showing the range of acceptable power supply voltages for computer equipment. The horizontal axis shows the duration of the sag or swell, and the vertical axis shows the percent change in line voltage.

In addition, the IEEE has addressed sag susceptibility and the economics of sag-induced events in IEEE Std. 1346–1998 [2.6]. This document includes measured power quality data taken from numerous sites.

In the 1990s, the Information Technology Industry Council (ITIC) curve was developed [2.7] by a working group of CBEMA. In recent years, the ITIC curve (Figure 2.6) has replaced the CBEMA curve in general usage for single-phase, 120-V, 60-Hz systems. A similar curve has been proposed for semiconductor processing equipment: the SEMI F47 curve. A comparison of these three curves is shown in Figure 2.7 [2.8].

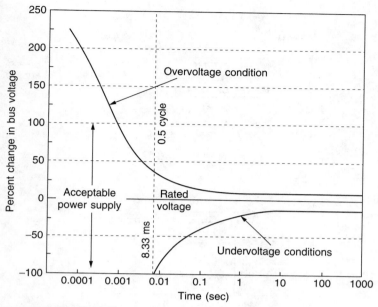

Figure 2.5 The CBEMA curve [2.5].
[© 2004, IEEE, reprinted with permission]

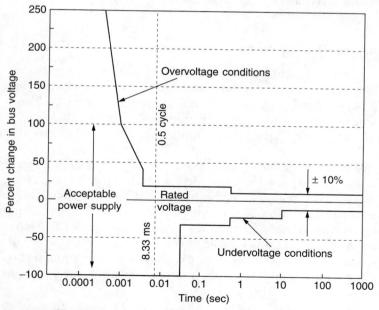

Figure 2.6 The ITIC curve [2.5].
[© 2004, IEEE, reprinted with permission]

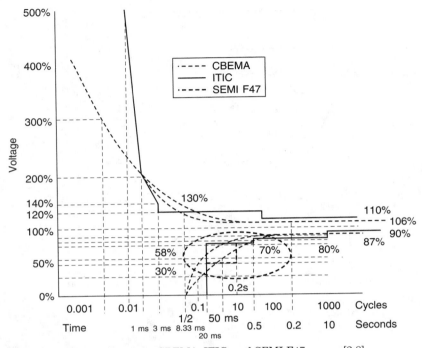

Figure 2.7   Comparison of the CBEMA, ITIC, and SEMI F47 curves [2.8].
[© 2005, IEEE, reprinted with permission]

High-Frequency EMI Standards

EMI standards relate to the design and testing of high-frequency switching power supply designs. There are limits to the amount of harmonic pollution a power supply is allowed to inject onto the power line. These limits depend on the frequency of operation, and the power level of the power supply used. Switching power supplies are discussed extensively in Chapter 7.

Shown in Figure 2.8 is a switching power supply[1] that draws power from an AC line. The AC line voltage is rectified and a large bus capacitor, C_{BUS}, creates a bus voltage. The bus capacitor voltage will have 120 Hz ripple due to the operation of the full-wave rectifier. The switching supply then chops the bus voltage at a very high frequency (high, that is, compared to the 60-Hz line frequency).

The line current i_s contains harmonics of the 60-Hz line frequency, as well as high-frequency interharmonics from the switching power supply.

---

[1] The high frequency switching supply block can be, for instance, a DC/DC converter or an adjustable speed drive.

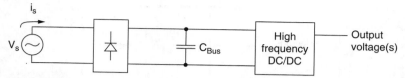

i$_s$

V$_s$ C$_{Bus}$ High frequency DC/DC   Output voltage(s)

Figure 2.8 A switching power supply that draws high frequency components from the AC power line.

Through design combinations of switching methods and EMI filtering, we can reduce but never completely eliminate the high frequencies injected into the AC line. These harmonics injected into the AC line are sometimes called "conducted emissions." Another effect of high-frequency harmonics injected onto the AC line is that the AC line will now radiate electromagnetic interference.

Another implementation that generates high-frequency harmonics on the line is the boost converter power factor correction circuit (Figure 2.9). This circuit is used in many high-power converters in the front end. This circuit draws high power factor current from the line, but the high-frequency switching of the MOSFET generates harmonics drawn from the line as well. The typical spectrum of the line current waveform for a DC/DC converter is shown in Figure 2.10.

The Federal Communications Commission (FCC), in their Rules, subpart J, sets limits for the conducted emissions allowable on power lines injected from line-connected equipment [2.10]. Class A covers industrial equipment, and class B covers residential equipment. Shown in Figure 2.11 is the FCC standard, which sets limits on the conducted noise injected onto the AC line in the 450 kHz to 30 MHz range. The Canadian agency CSA has similar limits as the FCC.

Other agencies regulating conducted EMI are the International Electrotechnical Commission (IEC) and the International Special Committee on Radio Interference (CISPR) [2.11]. CISPR has no regulatory authority but has been adopted by most European countries.

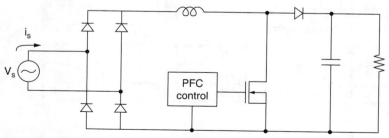

i$_s$

V$_s$ PFC control

Figure 2.9   Boost converter power factor correction circuit.

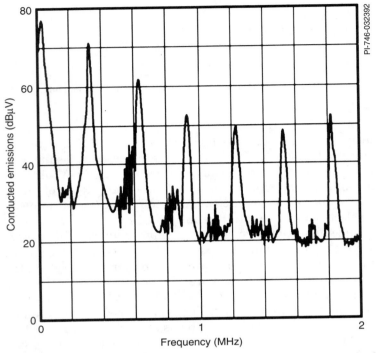

Figure 2.10 Typical conducted EMI spectrum from a DC/DC converter [2.9]. [© 1996, Power Integrations, reprinted with permission]

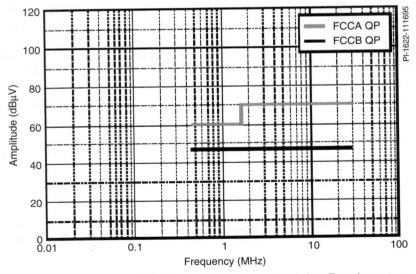

Figure 2.11 FCC-conducted EMI limits [2.9] for class A and class B equipment. [© 1996, Power Integrations, reprinted with permission]

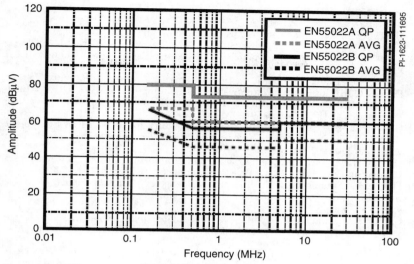

Figure 2.12   EN55022 conducted EMI limits [2.9].
[© 1996, Power Integrations, reprinted with permission]

One widely used standard in the European Community (EC) is the EN55022 standard [2.12] (Figure 2.12), which is based on the requirements set forth by CISPR [2.13] and covers the frequency range from 150 kHz to 30 MHz.

Other agencies such as the German VDE in document VDE 0871 set requirements for the German market. When applying high-frequency switching supplies, one must be mindful of the various limits set forth by the regulating agencies [2.14 and 2.15]. For a power supply to comply with these limits, the peak of the spectral lines must fall below specified limits. In addition to conducted EMI, CISPR and the FCC mandate limits on the radiated noise emitted from power supplies as well.

Summary

Power-quality standards address limits to harmonics and power-quality events at the point of common coupling in power systems. In this chapter, we have covered standards addressing 60-Hz harmonics (most notably, IEEE Std. 519 and 1159) as well as high-frequency standards that address harmonics created by high-frequency switching power supplies. Other standards, such as the CBEMA and ITIC curves set acceptable limits for sag and swell durations for computers and other information technology equipment.

References

[2.1] IEEE, "IEEE Recommended Practices and Requirements for Harmonic Control in Electrical Power Systems," IEEE Std. 519-1992, revision of IEEE Std. 519-1981.

[2.2] ____, "IEEE Recommended Practice for Monitoring Electric Power Quality," IEEE Std. 1159-1995.

[2.3] American National Standards Institute, "American National Standard Voltage Ratings (60Hz) for Electric Power Systems and Equipment," ANSI Std. C84.1-1989.

[2.4] IEEE, "IEEE Guide for Service to Equipment Sensitive to Momentary Voltage Disturbances," IEEE Std. 1250–1995.

[2.5] G. Lee, M. Albu, and G. Heydt, "A Power Quality Index Based on Equipment Sensitivity, Cost, and Network Vulnerability," *IEEE Transactions on Power Delivery*, vol 19, no 3, July 2004, pp. 1504–1510.

[2.6] IEEE, "IEEE Recommended Practice for Evaluating Electric Power System Compatibility with Electronic Process Equipment," IEEE Std. 1346-1998.

[2.7] ITIC curve is published by the Information Technology Industry Council, 1250 Eye St. NW, Suite 200, Washington D.C., 20005, or available on the Web at www.itic.com.

[2.8] S. Djokic, G. Vanalme, J. V. Milanovic, and K. Stockman, "Sensitivity of Personal Computers to Voltage Sags and Short Interruptions," *IEEE Transactions on Power Delivery*, vol. 20, no. 1, January 2005, pp. 375–383.

[2.9] Power Integrations, Inc., "Techniques for EMI and Safety," Application Note AN-15, June 1996; available from the Web: www.powerint.com.

[2.10] Code of Federal Regulations, Title 47, Part 15, Subpart J, "Computing Devices".

[2.11] CISPR, Publication 22, "Limits and Methods of Measurements of Radio Interference Characteristics of Information Technology Equipment," 1985.

[2.12] European Standard EN55022, "Limits and Methods of Measurement of Radio Interference Characteristics of Information Technology Equipment," CENELEC, 1994.

[2.13] R. Calcavecchio, "Development of CISPR 22 and Second Edition," *IEE Colloquium on Development of EMC Standards for Information Technology Equipment*, March 25, 1992, pp. 2/1–2/8.

[2.14] T. Curatolo and S. Cogger, "Enhancing a Power Supply to Ensure EMI Compliance," *EDN*, February 17, 2005, pp. 67–74.

[2.15] V. K. Dhar, "Conducted EMI Analysis—A Case Study," *Proceedings of the International Conference on Electromagnetic Interference and Compatibility '99*, December 6–8, 1999, pp. 181–186.

Voltage Distortion

In this chapter, we shall discuss distortion of the line-voltage sine waves. Distortion to the line-voltage waveforms can be caused by transient or continuous disturbances. Examples of transient disturbances include lightning, motor starting and stopping, clearing faults, and other harmonic-generating occurrences. Throughout the chapter, we use PSPICE simulation results to illustrate examples of voltage distortion. A chart summarizing voltage distortion and its causes is shown at the end of this chapter.

Voltage Sag

A *voltage sag*[1] is an event where the line rms voltage decreases from the nominal line-voltage for a short period of time. Figure 3.1 shows an 80 percent sag with a duration of a few 60-Hz cycles. This type of variation can occur if a large load on the line experiences a line-to-ground fault, such as a short in a three-phase motor or a fault in a utility or plant feeder.

In Figure 3.2a, we see a circuit with a line supplying an electric motor. Note that the line impedances cause a voltage drop when currents are drawn from the line. When the motor is energized, the motor current I_m causes a voltage drop to other loads in the system at the point of common coupling (PCC). Figure 3.2b displays a voltage sag due to a large motor starting, such as a pump or air-conditioner motor. Note that when an induction motor starts, it can draw very high currents until the rotor

---

[1] The IEEE term *sag* is a synonym to the IEC term *dip*.

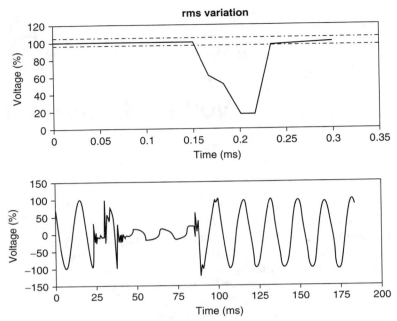

Figure 3.1 Voltage sag due to a single[2] line-to-ground fault [3.1].
[© 1995, IEEE, reprinted with permission]

comes up to speed.[3] This high current causes a significant voltage drop due to the impedance of the line.

Results from an EPRI study (Figure 3.3) [3.2] show that the average sag duration in U.S. systems is a few cycles long. This average dip clearing time of roughly 6 cycles (100 milliseconds) is attributable to the circuit breaker switching time for clearing a fault.

Example 3.1: Motor starting.   Next, we'll examine a hypothetical case of a three-phase motor starting and its effect on the load voltage. We'll model the system as shown in Figure 3.2a, where we have a 277-V line-neutral voltage, a resistance of 0.02 ohms, and an inductance of 100 microhenries. The resistance and inductance model the impedances of the combination of the utility, any wiring impedance, and the impedance of a step-down transformer The motor is modeled as a current source of 1000 A, which energizes at time $t = 0.1$ second. In Figure 3.4b, we see the effect of the motor starting on the load voltage, where there is a sag with approximately a 60-V peak. This corresponds to a sag of roughly 15 percent.

---

[2] From IEEE Std. 1159-1995.

[3] During startup, an induction motor will typically draw five to ten times the nominal full-load operating current.

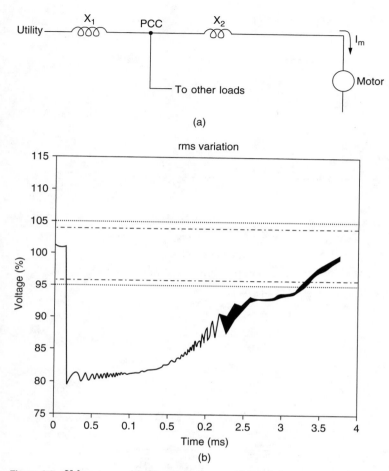

Figure 3.2 Voltage sag. (a) Circuit that can exhibit voltage sag due to motor starting. (b) Voltage sag caused by motor starting[4] [3.1]. [© 1995, IEEE, reprinted with permission]

Example 3.2: Voltage sag analysis. We shall next do a voltage sag analysis on a 480-V line-line system. The system is modeled as shown in Figure 3.5, with a line-neutral equivalent circuit. With a line-line voltage of 480 V, the source presents a line-neutral voltage of 277 V. The line reactance is (0.002 Ω + j0.01 Ω) and the load current is $I_L = 5000$ amperes. We'll assume the load current starts at time $t = 0.1$ seconds; the voltage drop across the line is thus:

$$V_{\text{drop}} = I_L (R + jX_L) = (5000)(0.002 + j0.01) = 10 + j(50)$$

---

[4] From IEEE Std. 1159-1995, [3.1], p. 18.

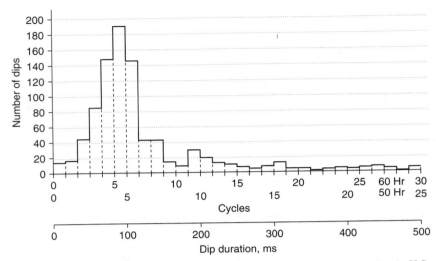

Figure 3.3 Histogram of voltage sag duration from EPRI study on voltage dips in U.S. Systems [3.2].
[© 2000, IEEE, reprinted with permission]

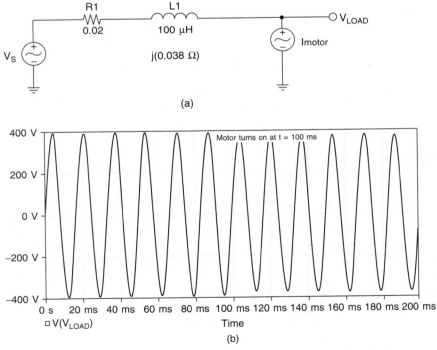

Figure 3.4 Voltage sag due to motor starting. (a) Line-neutral equivalent circuit, assuming 277-V line-neutral and a motor current of 1000 A rms. The motor current starts at $t = 100$ ms, and draws 1000 A at 60 Hz. (b) Voltage sag caused by motor starting.

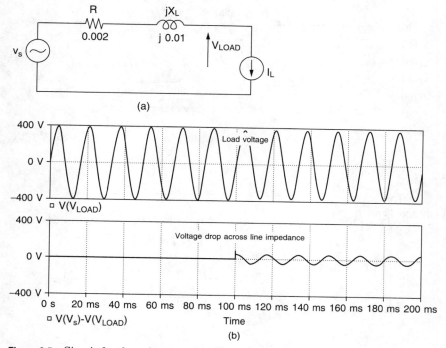

(a)

(b)

Figure 3.5 Circuit for the voltage sag analysis of Example 3.2. (a) Line-neutral equivalent circuit. (b) PSPICE simulation showing load current starting at $t = 0.1$ seconds. Top trace is line-neutral voltage with peak value 391.7 V (277 V rms). The bottom trace is the voltage drop across the line impedance.

The output voltage is $V_{\text{LOAD}} = 267 - \text{j}50$ V. We see that the inductive reactance results in a phase shift of:

$$\angle = -\tan^{-1}\left(\frac{50}{267}\right) = -10.6°$$

That is, the load voltage V_{LOAD} lags the source voltage v_s by 10.6°. The rms value of the voltage drop across the line impedance is 51 V.

A capacitor can be added on the load end to help the power factor, as shown in Figure 3.6. The reactive power provided by the added capacitor $(-\text{j}X_c)$ can improve the power factor.

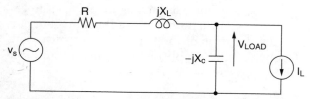

Figure 3.6 Adding a capacitor to offset the effects of a line-voltage drop.

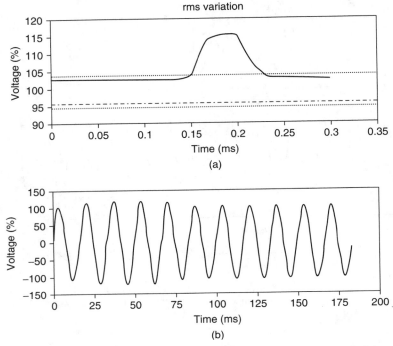

Figure 3.7 Voltage swell.[5] The top trace is the instantaneous rms value of the voltage. The bottom trace is the line-voltage [3.1].
[© 1995, IEEE, reprinted with permission]

Voltage "Swell"

A "swell" is the converse of the sag, and is a brief increase in the rms line-voltage. Shown in Figure 3.7 is a voltage swell caused by a line-to-ground fault.

Impulsive "Transient"

An impulsive "transient" is a brief, unidirectional variation in voltage, current, or both on a power line. The most common sources of impulsive transients are lightning strikes (Figure 3.8). Impulsive transients due to lightning strikes[6] can occur because of a direct strike to a power line, or from magnetic induction or capacitive coupling from strikes on adjacent lines.

Note that in this case, the maximum amplitude of the transient current is ~23 kiloamps and the duration of the transient current is tens of microseconds. The frequency and amplitude of lightning-induced transients vary geographically, as shown in Figure 3.9.

---

[5] From IEEE Std. 1159-1995, p. 19.

[6] Also called lightning "strokes."

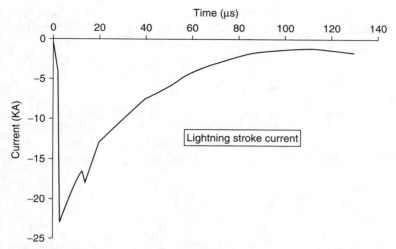

Figure 3-8   Impulsive transient[7] [3.1].
[© 1995, IEEE, reprinted with permission]

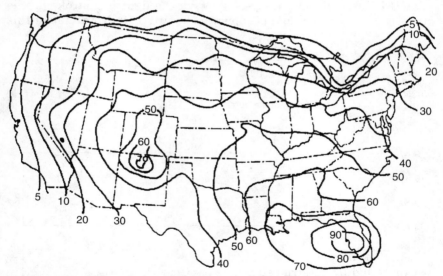

Figure 3.9   Number of thunderstorm days per year[8] in the U.S. [3.3].
[© 1995, IEEE, reprinted with permission]

Example 3.3: Voltage impulse due to a lightning strike.   In this example, we'll calculate the instantaneous voltage drop across the line impedance, assuming the lightning stroke profile of Figure 3.8 and the line

---

[7] From IEEE Std. 1159-1995, p. 13.

[8] From IEEE Std. 446-1995, p. 7.

impedance from Example 3.2. Note that the line impedance is 0.002 Ω + j0.01 Ω. At high frequencies, the line impedance is dominated by the line inductance. In this case, the inductance is

$$L = \frac{X_L}{2\pi f} = \frac{X_L}{2\pi(60)} = \frac{0.01}{2\pi(60)} = 26.5 \ \mu H$$

We'll approximate the negative rising edge of the lightning strike as rising to −23 kiloamperes in 5 microseconds. The lightning strike then decays to zero in approximately 100 microseconds. Using these assumptions, and remembering that $v = L di/dt$ and ignoring the voltage drop across the resistor, we find

$$v_1 = (26.5 \times 10^{-6})\left(\frac{-23,000}{5 \times 10^{-6}}\right) = -121.9 \ kV$$

$$v_2 = (26.5 \times 10^{-6})\left(\frac{23,000}{100 \times 10^{-6}}\right) = +6.1 \ kV$$

A PSPICE circuit and analysis showing the lightning stroke profile and effects on load voltage is shown in Figure 3.10. Note that in this

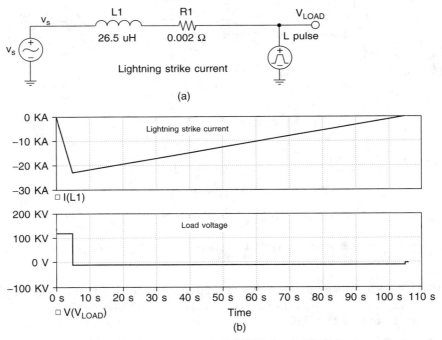

Figure 3.10 Simulation for Example 3.3. (a) A PSPICE circuit. The lightning strike is modeled as a triangular pulse of current injected onto the line. (b) PSPICE result showing lightning strike current and load voltage.

example, we assume there are no lightning arresters or insulation break-down to limit the transient voltages.

Oscillatory "Transient"

An oscillatory transient is a brief, bidirectional variation in voltage, current, or both on a power line. These transients can occur due to res-onances during switching. A circuit capable of exhibiting this phenom-enon is shown in Figure 3.11a. A power supply bus is shown with the bus having inductance L. A capacitor bank labeled C_1 is connected at one end of the bus. This capacitor bank may be in place, for instance, for power factor improvement or for voltage sag improvement.

If at some time, we switch in capacitor bank C_2 with the switch as shown, a resonant condition is set up between the line inductance and the

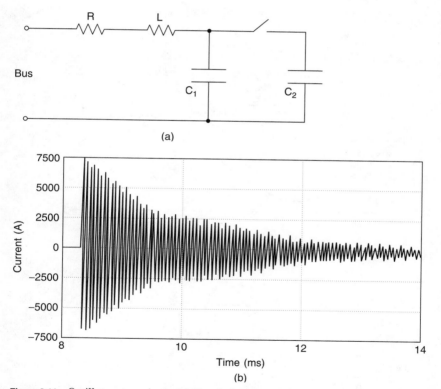

(a)

(b)

Figure 3.11 Oscillatory transients. (a) Circuit that can produce an oscillatory transient. (b) An oscillatory transient[9] [3.1].
[© 1995, IEEE, reprinted with permission]

---

[9] From IEEE Std. 1159-1995, p. 14.

capacitor banks. The resultant resonance will be underdamped, and the current in the capacitor bank may look something like that in Figure 3.11b.

Example 3.4: Capacitor bank switching.   We'll next consider an example of capacitor bank switching. Let's look at the circuit of Figure 3.12a, where we have a line with a capacitor bank at the load end of the line. The capacitor bank may be for power factor correction, or for some other reason. In this simulation, the line-neutral voltage is 277 V and the capacitor bank is switched in at $t = 100$ milliseconds. We see in Figure 3.12b the resultant ripple in the load voltage and capacitor current after the

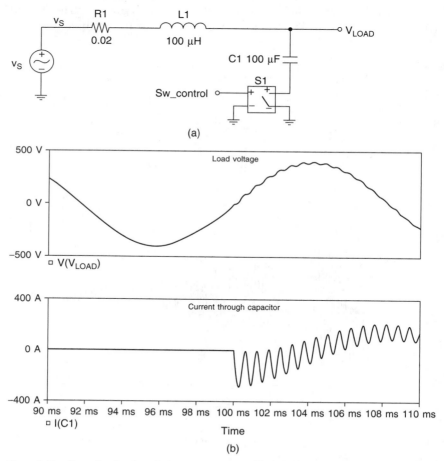

Figure 3.12   Capacitor bank switching. (a) Circuit. (b) PSPICE simulation results. The top trace is the instantaneous line-voltage, while the bottom trace is the current in the capacitor. The capacitor is switched into the circuit at $t = 100$ ms, and the capacitor current rings, resulting in a ripple in the load voltage.

capacitor bank is switched. The ripple in the load voltage is due to the capacitor current that rings at the resonant frequency of the LC circuit.[10]

Interruption

An interruption is defined as a reduction in line-voltage or current to less than 10 percent of nominal, not exceeding 60 seconds in length. Interruptions can be a result of control malfunction, faults, or improper breaker tripping. Figure 3.13 shows an interruption of approximately 1.7 seconds in length.

Notching

Another common power-quality event is "notching." Notching can occur during current "commutation" in single-phase and three-phase rectifiers. Shown in Figure 3.14a is a three-phase rectifier with the line inductance L_s of each phase shown. Note that in the rectifier, if we assume that the

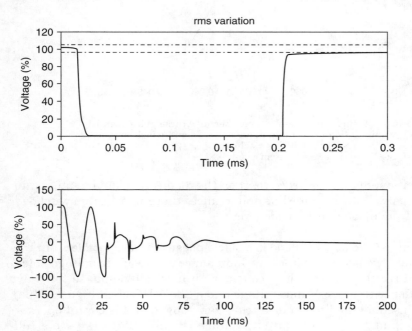

Figure 3.13 An interruption of approximately 0.17 seconds duration[11] [3.1]. The top trace is the rms line-voltage. The bottom trace is the first 200 milliseconds of the interruption.
[© 1995, IEEE, reprinted with permission]

---

[10] The ring frequency is $1/2\pi\sqrt{LC}$ radians/second, or 1591 Hz.

[11] From IEEE Std. 1159-1995, p. 16.

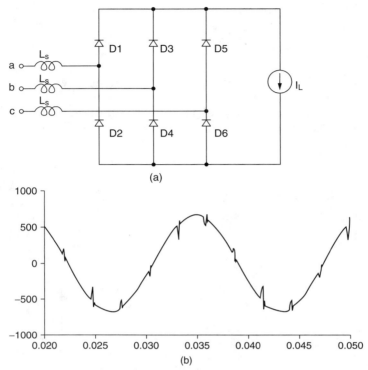

(a)

(b)

Figure 3.14   Notching in a three-phase rectifier. (a) A three-phase rectifier that has commutation due to line inductances L_s and which produces notching. (b) A waveform[12] showing notching.
[© 1995, IEEE, reprinted with permission]

parasitic inductances are zero, then the diodes turn on and off instantaneously. With a finite line inductance, there is a finite switchover time from diode pair to diode pair.

Example 3.5: Voltage notching in a single-phase full-wave rectifier.   In this example, we'll see how line-side inductance can result in voltage notching in rectifiers. First, let's consider an ideal single-phase full-wave rectifier with a 208-V line-voltage and a 200 A DC load (Figure 3.15a). If we assume the line inductance is zero, how will this circuit operate? Looking at the rectifier output voltage (Figure 3.15b), we see a full-wave rectified sine wave, as expected. In this circuit, $D1$ and $D4$ are on for the positive half-wave of the sine wave, and $D2$ and $D3$ are on for the negative half-wave. At the sine wave zero crossing, the switchover from diode pair to diode pair occurs instantaneously.

---

[12] From IEEE Std. 1159-1995, p. 23.

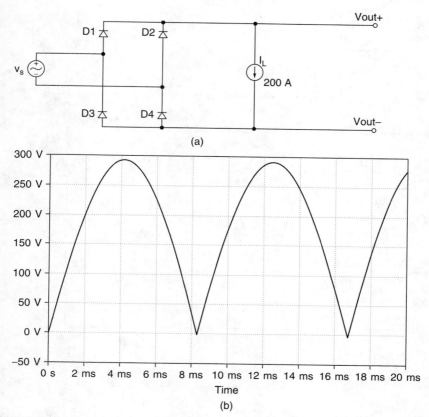

Figure 3.15 An ideal single-phase full-wave rectifier with a 208-V line-neutral voltage and loaded by a constant current of 200 A DC. (a) Circuit; (b) output voltage waveform. There is no notching, since in this exercise we've assumed the line inductance is negligible.

Next, let's consider what happens if there is a finite line inductance. In Figure 3.16a, we see the same circuit, but this time assuming there is 100 microhenries of line inductance. The effect of the inductor is to introduce a finite switchover time from one pair of diodes turning off to the other pair turning on. During this switchover time, all four diodes are on and the output of the rectifier is zero. This effect is shown in Figure 3.16b where we see notches near the sine wave zero crossing. This notching adds undesirable harmonics to the load voltage, and also reduces the average value of the load voltage.

Voltage Fluctuations and Flicker

Voltage fluctuations are relatively small (less than ±5 percent) variations in the rms line-voltage. Shown in Figure 3.17 is a system setup

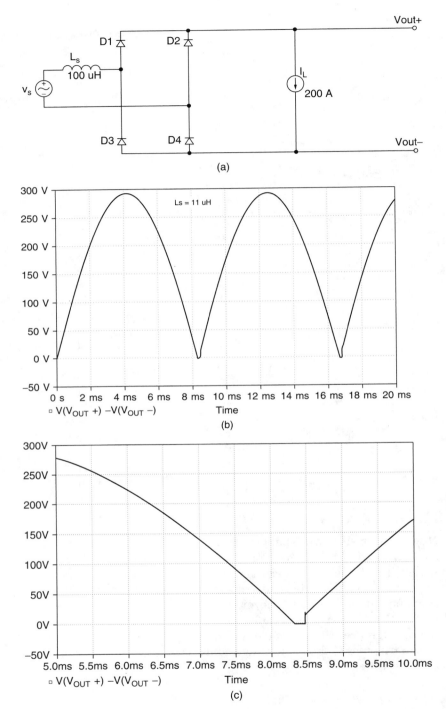

Figure 3.16 A nonideal single-phase full-wave rectifier with finite line inductance. (a) Circuit; (b) the output voltage waveform. Note the notching in the output voltage waveform. (c) Close-up of the waveform in the 5ms to 10ms time range showing the notching in the full-wave rectified wave.

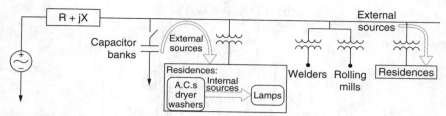

Figure 3.17   A circuit capable of flicker propagation to a residence [3.4].
[© 2004, IEEE, reprinted with permission]

that can cause voltage variations. A varying source of harmonic currents includes welders and capacitor banks. Voltage variation created by this setup couples to residential lighting through the distribution system.

Figure 3.18 displays typical harmonic levels during arc furnace operation. The harmonics produced by an arc furnace are unpredictable due to the variation of the arc during metal melting. We see that during initial melting, the harmonic content (both even and odd harmonics of the line-voltage) are relatively high. During the latter part of the arc furnace melt cycle, the arc is more stable and the harmonic current has diminished. Shown in Figure 3.19 is an example of line-voltage fluctuations caused by the operation of an arc furnace. Such voltage fluctuations can cause a further phenomenon known as "flicker." The waveform of a flicker event is shown in Figure 3.19.

Flicker is the human perception of light intensity variation. In Figure 3.20, we see a "flicker curve," which shows that the human perception of flicker depends on the amplitude and frequency of the event.

Table 4.1
Harmonic Content of Arc Furnace Current
at Two Stages of the Melting Cycle

	Harmonic Current % of Fundamental				
	Harmonic Order				
Furnace condition	2	3	4	5	7
Initial melting (active arc)	7.7	5.8	2.5	4.2	3.1
Refining (stable arc)	0.0	2.0	0.0	2.1	0.0

Figure 3.18   Harmonic content of arc furnace current [3.5].
[© 1992, IEEE, reprinted with permission]

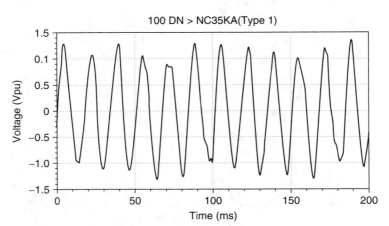

Figure 3.19   Voltage fluctuations [3.1].
[© 1995, IEEE, reprinted with permission]

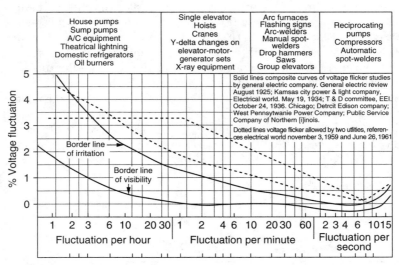

Figure 3.20   Voltage fluctuation limits, from [3.5], p. 81.
[© 1992, IEEE, reprinted with permission]

Voltage Imbalance

A voltage "imbalance" is a variation in the amplitudes of three-phase voltages, relative to one another. Figure 3.21 shows a three-phase voltage waveform where the phases a, b, and c have different amplitudes. This

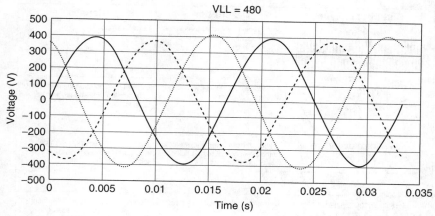

Figure 3.21   Voltage imbalance.

imbalance can be caused by different loads on the phases, resulting in different voltage drops through the phase-line impedances.

A voltage imbalance can cause a reverse-rotating airgap field in induction machines, increasing heat loss and temperature rise.

Summary

To summarize voltage distortion types and causes, see Figure 3.22.

Disturbance type	Description	Causes
Impulse	Narrow pulse with fast rise and exponential or damped oscillatory decay; 50 V to 6 kV amplitude, 0.5 μs to 2 ms duration	Load switching, fuse clearing, utility switching, arcing contacts, lightning
EMI	Repetitive low-energy disturbances in the 10 kHz to 1 GHz band, with 100 μV to 100 V amplitude	Normal equipment opeation (switching power supplies, motor speed controllers, etc.), carrier power-line communication, wireless broadcasting
SAG	Low voltage (typ. less than 80%), for more than one periord	Starting heavy load, utility switching, ground fault
Swell	High voltage (typ. more than 110%), for more than one period	Load reduction, utility switching

Figure 3.22   Overview of power disturbances [3.6].
[© 1995, IEEE, reprinted with permission]

Disturbance type	Description	Causes
Flicker	Small repetitive fluctuations in the voltage level	Pulsating load
Notches	Repetitive dips in the line voltage, with short durations	Current commutation in controlled or uncontrolled three-phase rectifier circuits
Waveform distortion	Deviation from ideal sine wave due to the presence of harmonics or interharmonics	Rectifiers, phase-angle controllers, other nonlinear and/or intermittent loads
Frequency variation	Deviation of the frequency from the nominal value	Poorly regulated utility equipment, emergency power generator
Outage	Zero-voltage condition of a single phase or several phases in a multi-phase system, for more than a half-period	Load equipment failure, ground fault, utility equipment failure, accidents, lightning, acts of nature

Figure 3.22   (*Continued*)

References

[3.1] IEEE, "IEEE Recommended Practice for Monitoring Electric Power Quality," IEEE Std. 1159-1995.

[3.2] K. J. Cornick and H. Q. Li, "Power Quality and Voltage Dips: Problems, Requirements, Responsibilities," *Proceedings of the 5th International Conference on Advances in Power System Control, Operation, and Management, APSCOM 2000*, Hong Kong, October 2000, pp. 149–156.

[3.3] IEEE, "IEEE Recommended Practice for Emergency and Standby Power Systems for Industrial and Commercial Applications," IEEE Std. 446-1995 (The Orange Book).

[3.4] C.-S. Wang and M. J. Devaney, "Incandescent Lamp Flicker Mitigation and Measurement," *IEEE Transactions on Instrumentation and Measurement*, vol. 53, no. 4, August 2004, pp. 1028–1034.

[3.5] IEEE, "IEEE Recommended Practices and Requirements for Harmonic Control in Electrical Power Systems," IEEE Std. 519-1992, revision of IEEE Std. 519-1981.

[3.6] R. Redl, R. and A. S. Kislovski, "Telecom Power Supplies and Power Quality," *Proceedings of the 17th International Telecommunications Energy Conference, INTELEC '95*, October 29 to November 1, 1995, pp. 13–21.

Harmonics and Interharmonics

In this chapter, we shall discuss harmonics (frequency components that are integer multiples of the fundamental line frequency) and interharmonics (high-frequency components). For most of what we shall do in this chapter, the fundamental frequency used will be 60 Hz.

Background

As we mentioned in previous chapters, harmonics can adversely affect the operation of cables, capacitors, metering, and protective relays. To summarize, a brief listing of some systems and the effects of harmonics is shown in Table 4.1 [4.1].

Periodic Waveforms and Harmonics

The notion that any periodic waveform can be broken up into a series of sine waves at the proper amplitudes and phase relationships was first worked out by Joseph Fourier, the French mathematician and physicist [4.2]. He showed that any periodic waveform can be expressed as a sum of sine and/or cosine waves, with the proper amplitude, frequency, and phase relationships between the waves. For instance, a square wave (Figure 4.1a) can be represented by the infinite Fourier series:

$$v(t) = \left(\frac{4}{\pi}\right)\sin(\omega t) + \left(\frac{4}{3\pi}\right)\sin(3\omega t) + \left(\frac{4}{5\pi}\right)\sin(5\omega t) + \cdots$$

where ω is the frequency in radians per second. Note that the amplitude of the first harmonic is $(4/\pi)$, the amplitude of the third harmonic is

TABLE 4.1   **Some Effects of Harmonics**

At	Effect
Circuit breakers	Malfunction
Capacitor banks	Overheating Insulation breakdown Failure of internal fuses
Protection equipment	False tripping No tripping
Measuring devices	Wrong measurements
Transformers, reactors	Overheating
Motors	Increased noise level Overheating Additional vibrations
Telephones	Noise with the respective harmonic frequency
Lines	Overheating
Electronic devices	Wrong pulses on data transmission Over-/undervoltage Flickering screens
Incandescent lamps	Reduced lifetime Flicker

one-third as much the first, the amplitude of the fifth harmonic is one-fifth as much as the first harmonic, and so on. Also, note that the square wave has only odd harmonics (that is, harmonics of the order 1, 3, 5..., and so on). The spectrum of the square wave is shown in Figure 4.1b.

Likewise, a triangle wave (Figure 4.2a) can be represented by the infinite Fourier series:

$$v(t) = \left(\frac{8}{\pi^2}\right)\sin(\omega t) - \left(\frac{8}{3^2\pi^2}\right)\sin(3\omega t) + \left(\frac{8}{5^2\pi^2}\right)\sin(5\omega t) + \cdots$$

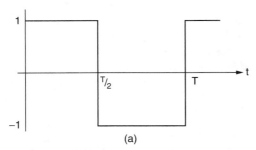

(a)

Figure 4.1 Periodic waveforms. (a) A square wave with peak values ±1 and period T. (b) The spectrum of a square wave. The square wave has only odd harmonics.

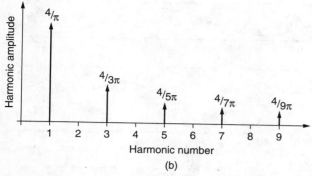

Figure 4.1 (*Continued*)

We see that the harmonics of the triangle wave (Figure 4.2b) fall off at a faster rate than those of a square wave. This makes sense since the triangle wave more closely resembles a pure sine wave than a square wave, and therefore has fewer harmonics than the square wave.

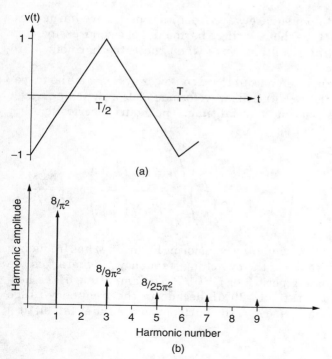

Figure 4.2 Periodic waveforms. (a) A triangle wave. (b) The spectrum of a triangle wave.

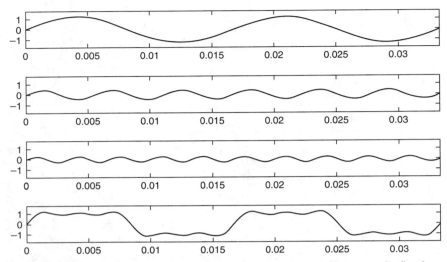

Figure 4.3 The first three harmonics that make up a square wave. Shown are the first harmonic at 60 Hz (top trace), third and fifth harmonics, and the total waveform (bottom trace) that is the sum of the three harmonics.

Next, we'll build up a square wave from its constitutive harmonics. Shown in Figure 4.3 are the first three harmonics of a square wave (top three traces) and the resultant wave when the three harmonics are added (bottom trace).

Another waveform often encountered in power systems is the trapezoidal waveform (Figure 4.4). This waveform models a switching waveform with a finite risetime and falltime. The Fourier series for this waveform is given by [4.3]:[1]

$$i_D(t) = \left(\frac{2T_D}{T}\right) \sum_{N=1,2,3\ldots}^{\infty} \left(\frac{\sin \pi N \left(\frac{T_D}{T}\right)}{\pi N \left(\frac{T_D}{T}\right)}\right) \left(\frac{\sin \pi N \left(\frac{t_r}{T}\right)}{\pi N \pi N \left(\frac{t_r}{T}\right)}\right) \cos\left(\frac{2\pi Nt}{T}\right)$$

The spectrum for this switching waveform (Figure 4.4) has frequency components at multiples of the switching frequency f_o, where f_o is the inverse of the switching period, or $f_o = 1/T$. The amplitude of the harmonics falls off at a rate of –20 dB/decade in the frequency range between f_1 and f_2, while above f_2 the harmonic amplitudes fall off at a

---

[1] This equation assumes the risetime and falltime of the trapezoid are the same.

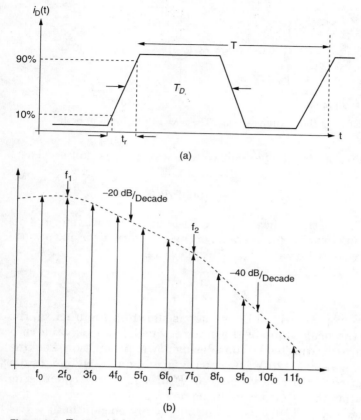

Figure 4.4 Trapezoidal waveform. (a) Current waveform that may be produced by a DC/DC converter or other electronic system. (b) Spectrum.

rate of 40 dB/decade. It can be shown that the two corner frequencies f_1 and f_2 are found by [4.4], [4.5]:

$$f_1 = \frac{1}{\pi T_d}$$

$$f_2 = \frac{1}{\pi t_r}$$

Root-mean square

Root-mean square is a measure of the heating value of a periodic waveform when this periodic waveform drives a resistive load. Mathematically, the

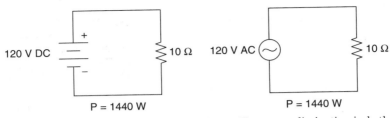

P = 1440 W   P = 1440 W

Figure 4.5 Illustration of the meaning of rms. The power dissipation in both cases is the same.

root-mean square or rms of a periodic voltage waveform is expressed as:

$$V_{rms} = \sqrt{\frac{1}{T}\int_0^T [v(t)]^2 dt}$$

where we see inside the radical that we first square the waveform, and then take the mean value (or average) of the waveform over one period. For a sine wave of peak value V_{pk}, the rms value is

$$V_{rms} = \frac{V_{pk}}{\sqrt{2}}$$

For a square wave (with no DC value) as shown in Figure 4.1a, the rms value is the peak value of the square wave. The rms value of a waveform can be interpreted by considering power dissipation. Looking at Figure 4.5, we see a 120-V DC battery driving a 10-Ω load, and a 120-V AC source (with rms value 120 V) driving a 10-Ω load. The power dissipation in both loads is the same at 1440 W.

In the following, we'll discuss a few commonly encountered waveforms in power systems and power electronics, and their corresponding rms (root-mean square) values [4.6]. Remember that the rms value of a periodic waveform is the square root of the average value of the *square* of the waveform over a period. For a periodic current i(t), the corresponding rms value is

$$I_{rms} = \sqrt{\frac{1}{T}\int_0^T i^2(t) dt}$$

i(t)

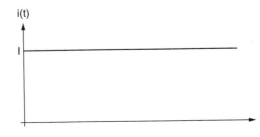

Figure 4.6   DC current.

t

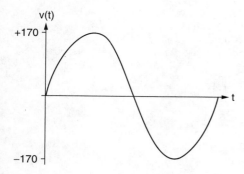

Figure 4.7 A pure sine wave.

DC current

A DC current (Figure 4.6) has an rms value equal to the steady-state current, or:

$$I_{\text{rms}} = I$$

Pure sine wave

A pure sine wave (Figure 4.7) has an rms value equal to the peak value divided by the square root of two. In the case of a sine wave with peaks at ±170 V, the rms value is

$$V_{\text{rms}} = \frac{V_{\text{pk}}}{\sqrt{2}} = \frac{170}{\sqrt{2}} = 120\,\text{V}$$

Square wave

A 50 percent square wave (Figure 4.8) can be generated by full-bridge and half-bridge power converters. The rms value of this waveform is

$$I_{\text{rms}} = I_{\text{pk}}$$

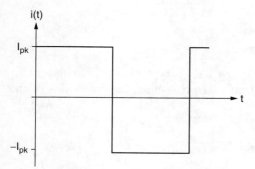

Figure 4.8 A 50 percent duty-cycle square wave.

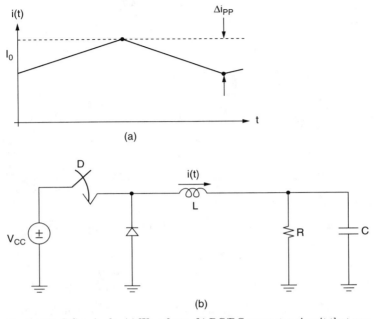

Figure 4.9   DC + ripple. (a) Waveform. (b) DC/DC converter circuit that produces this waveform. In this case, $i(t)$ is the inductor current.

DC waveform + ripple

A DC waveform with a finite peak-peak ripple (Figure 4.9) is generated by a variety of switching circuits, including motor drives and DC/DC converters. This waveform has a DC value I_o and a peak-peak ripple of Δi_{pp}. The rms value of this waveform is

$$I_{rms} = I_o\sqrt{1 + \left(\frac{1}{3}\right)\left(\frac{\Delta i_{pp}}{2I_o}\right)^2}$$

Triangular ripple

A triangular ripple waveform (Figure 4.10) is typical of the capacitor current in a variety of switching circuits. In this case, the rms value of the current is

$$I_{rms} = \frac{\Delta i_{pp}}{2\sqrt{3}}$$

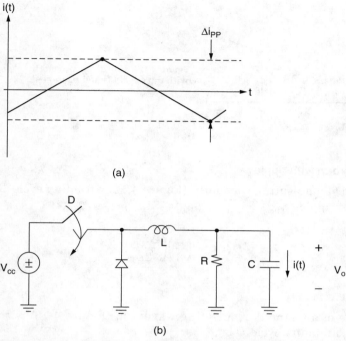

Figure 4.10   Triangular ripple. (a) A waveform. (b) The DC/DC converter circuit that produces this waveform. In this case, $i(t)$ is the capacitor current.

Pulsating waveform

The rms value of a pulsating waveform (Figure 4.11) with duty cycle D is

$$I_{\text{rms}} = I_{\text{pk}} \sqrt{D}$$

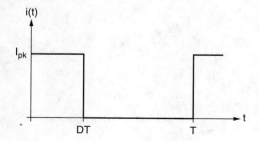

Figure 4.11   Pulsating waveform with duty cycle D.

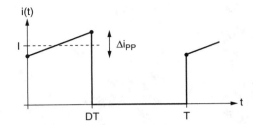

Figure 4.12 Pulsating waveform with duty cycle D and peak-peak ripple Δi_{pp}.

Pulsating waveform with ripple

The rms value of a pulsating waveform (Figure 4.12) with duty cycle D and peak-peak ripple Δi_{pp} is

$$I_{\mathrm{rms}} = I\sqrt{D}\sqrt{1 + \left(\frac{1}{3}\right)\left(\frac{\Delta i_{\mathrm{pp}}}{2I}\right)^2}$$

Triangular waveform

The rms value of a triangular pulsating waveform (Figure 4-13) with peak value I_{pk} and duty cycle D is

$$I_{\mathrm{rms}} = I_{\mathrm{pk}}\sqrt{\frac{D}{3}}$$

Piecewise Calculation

Let's assume we have a periodic waveform $i(t)$ that can be broken up into different frequency components $I_1, I_2, I_3, \ldots$ etc. The rms value of I_1 is $I_{1,\mathrm{RMS}}$. The rms value of the total waveform made up of the sum of

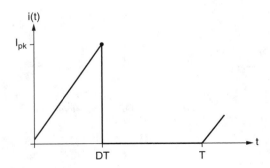

Figure 4.13 A triangular waveform.

individual currents is the sum of the squares of the rms values of the individual components, or:

$$I_{rms} = \sqrt{I_{1,rms}^2 + I_{2,rms}^2 + I_{3,rms}^2 + \cdots}$$

Total Harmonic Distortion

Total harmonic distortion or *THD* is a measure of how much harmonic content there is in a waveform. The total harmonic distortion of a waveform is

$$THD = \sqrt{\frac{V_{rms}^2 - V_{1,rms}^2}{V_{1,rms}^2}}$$

where V_{RMS} is the rms value of the total waveform, and $V_{1,RMS}$ is the rms value of the first harmonic. The THD of a sine wave is 0 percent, and the THD of a square wave is 48 percent.

Crest Factor

Crest factor is another term sometimes used in power systems analysis, and represents the ratio of the peak value to the rms value of a waveform. For a sine wave (Figure 4.14a), the peak value is 1.0 and the rms value is 0.707. Thus, the crest factor is 1.414. For a square wave (Figure 4.14b), the peak and rms values are both 1.0—hence, the crest factor is 1.0.

Example 4.1: A truncated square wave.   A square wave with peak value +1 has the Fourier series:

$$v(t) = \sum_{n=1}^{\infty} \left(\frac{4}{n\pi}\right) \sin\left(2\pi nt\right)$$

A truncated Fourier series approaches the ideal waveform. We'll next find the rms value and total harmonic distortion for a square wave waveform with harmonics present up to the seventh. The total waveform for this example is

$$v(t) = \left(\frac{4}{\pi}\right)\sin(\omega t) + \left(\frac{4}{3\pi}\right)\sin(3\omega t) + \left(\frac{4}{5\pi}\right)\sin(5\omega t) + \left(\frac{4}{7\pi}\right)\sin(7\omega t)$$

The rms value of the first harmonic is

$$V_{1,rms} = \frac{4}{\pi\sqrt{2}} = 0.9$$

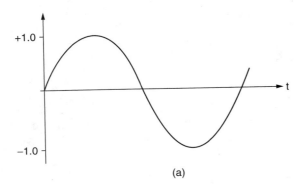

(a)

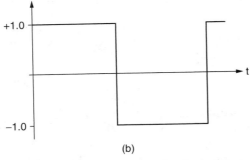

(b)

Figure 4.14 Waveforms illustrating the crest factor. (a) A sine wave with peak value 1.0. (b) A square wave with peak value 1.0.

The rms value of this total waveform is

$$V_{rms} = \sqrt{\left(\frac{4}{\pi\sqrt{2}}\right)^2 + \left(\frac{4}{3\pi\sqrt{2}}\right)^2 + \left(\frac{4}{5\pi\sqrt{2}}\right)^2 + \left(\frac{4}{7\pi\sqrt{2}}\right)^2}$$

$$= \left(\frac{4}{\pi\sqrt{2}}\right)\sqrt{1 + \left(\frac{1}{3}\right)^2 + \left(\frac{1}{5}\right)^2 + \left(\frac{1}{7}\right)^2} = 0.974$$

The total harmonic distortion is

$$\text{THD} = \sqrt{\frac{V_{rms}^2 - V_{1,rms}^2}{V_{1,rms}^2}} = \sqrt{\frac{(0.974)^2 - (0.9)^2}{(0.9)^2}} = 0.414 = 41.4\%$$

This waveform is shown in Figure 4.15.

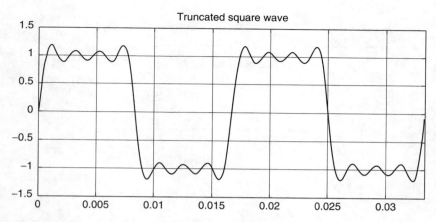

Figure 4.15 A truncated square wave Fourier series result with first, third, fifth, and seventh harmonics. The THD is 41.4 percent.

Example 4.2: Neutral current in three-phase systems. In this example, we'll show that the neutral current in three-phase systems with balanced linear loads is zero. Shown in Figure 4.16a is a balanced three-phase system with linear resistive loads. The three-phase voltages have the form:

$$v_{an} = V\sin(\omega t)$$

$$v_{bn} = V\sin(\omega t - 120°)$$

$$v_{cn} = V\sin(\omega t - 240°)$$

where each of the voltages (v_{an}, v_{bn}, v_{cn}) is the phase voltage to neutral. The phase currents are:

$$i_a = \left(\frac{V}{R}\right)\sin(\omega t)$$

$$i_b = \left(\frac{V}{R}\right)\sin(\omega t - 120°)$$

$$i_c = \left(\frac{V}{R}\right)\sin(\omega t - 240°)$$

The neutral current is the vector sum of the three-phase currents.

$$i_n = i_a + i_b + i_c = \left(\frac{V}{R}\right)[\sin(\omega t) + \sin(\omega t - 120°) + \sin(\omega t - 240°)] = 0$$

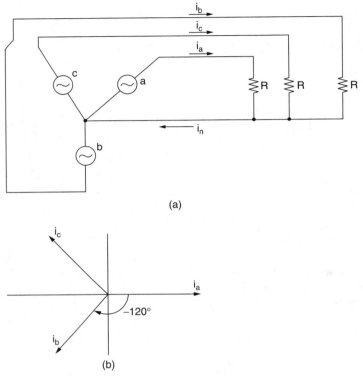

(a)

(b)

Figure 4.16 A balanced three-phase system. (a) A circuit showing balanced resistive loads. (b) A phasor diagram of the neutral currents.

Referring to the phase current phasor diagram (Figure 4.16b), note that the neutral current indeed does sum to zero in this special case of balanced, linear loads.

Example 4.3: Nonlinear loads. Power-line harmonics are created when nonlinear loads draw nonsinusoidal current from a sinusoidal voltage source. In this example, we'll show how nonlinear loads can result in high neutral currents in three-phase systems. These harmonics can result in neutral current that exceeds the individual phase current.

Let's consider a three-phase system where the loads on each of the three phases are balanced but nonlinear. Therefore, the magnitudes of the current in each phase are equal to one another. In most three-phase systems with nonlinear loads, the odd harmonics dominate the even harmonics. Mathematically, we can express the current in phases a, b, and c as:

$$i_a = I_1 \sin(\omega t - \theta_1) + \sum_{n=2j+1}^{\infty} I_n \sin(n\omega t - \theta_n)$$

$$i_b = I_1 \sin(\omega t - \theta_1 - 120°) + \sum_{n=2j+1}^{\infty} I_n \sin(n\omega t - \theta_n - n120°)$$

$$i_c = I_1 \sin(\omega t - \theta_1 - 240°) + \sum_{n=2j+1}^{\infty} I_n \sin(n\omega t - \theta_n - n240°)$$

with $j = 1, 2, 3, \ldots$ We see that I_1 is the amplitude of the fundamental, and the I_ns are the amplitudes of the odd harmonics. There are phase shifts, denoted by θ_n, for each of the harmonics as well.

To further simplify matters, let's next consider a load that produces only third-harmonic currents. In many three-phase circuits, the third harmonic is the dominant harmonic. In three-phase systems where there are third-harmonic currents, the 120-degree phase shift for the fundamental results in a 360-degree phase shift for the third harmonic. This means that the third-harmonic currents from each phase conductor are in phase with one another, and that the neutral current is equal to the sum of the third-harmonic amplitudes from each of the phases, or:

$$i_n = 3I_{h3}$$

where I_{h3} is the amplitude of the third-harmonic current in each phase. We'll next examine this phenomenon graphically, using a system that has a third-harmonic amplitude on each phase that is 30 percent of the fundamental. Shown in Figure 4.17 (top traces) are the first and third harmonics of this phase conductor. Note that the peak of the 60 Hz fundamental is 1.0 amps, and the peak of the 180 Hz third harmonic is 0.3 amps. Shown in the bottom trace of Figure 4.17 is the total phase-a current, which is the vector sum of the fundamental and third harmonic.

Next, we add up the sum of the phase currents to get the total neutral current.

$$i_a = (1.0) \sin(\omega t) + (0.3) \sin(3\omega t)$$

$$i_b = (1.0) \sin(\omega t - 120°) + (0.3) \sin(3\omega t - (3) \times 120°)$$

$$i_c = (1.0) \sin(\omega t - 240°) + (0.3) \sin(3\omega t - (3) \times 240°)$$

$$i_n = i_a + i_b + i_c = 0.9 \sin(3\omega t)$$

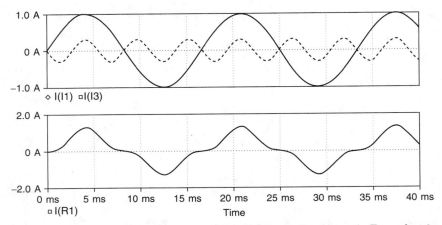

Figure 4.17 A plot of the fundamental and third harmonic for phase a in Example 4.3. The top traces show the fundamental and third harmonic currents. The bottom trace is the vector sum.

Note that the vector sum of the fundamental of the neutral current is zero. The vector sum of the third-harmonic neutral current is three times that of the third-harmonic amplitude of each phase.

Example 4.4: Total harmonic distortion. In this example, we'll find the total harmonic distortion for a voltage waveform with the harmonic amplitudes shown in Table 4.2. The THD for this waveform is found simply by:

$$\text{THD} = \frac{\sqrt{0.25^2 + 0.15^2 + 0.10^2 + 0.08^2 + 0.05^2 + 0.04^2}}{1.0} = 33.5\%$$

Example 4.5: Effects of load current harmonics on load voltage and THD. In the next example, we'll see the effects of load current harmonics on load voltage and total harmonic distortion. Figure 4.18 presents a system

TABLE 4.2 The Harmonic Voltage Spectrum for Example 4.5

Harmonic number	Amplitude
1	1.0
5	0.25
7	0.15
11	0.10
13	0.08
17	0.05
19	0.04

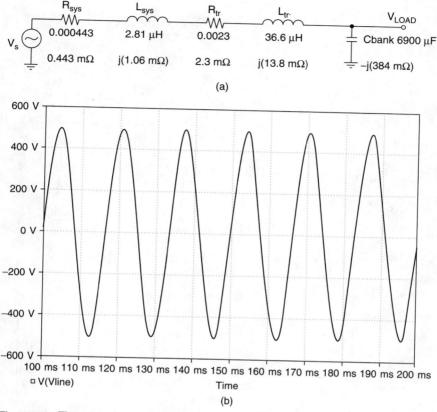

Figure 4.18 The original system for Example 4.5. (a) A circuit for PSPICE analysis, showing component values as well as the reactance of each component at 60 Hz. (b) Load voltage.

modeled as a voltage source in series with a source impedance (R_{sys}, L_{sys}) and a transformer impedance (R_{tr}, L_{tr}). We've added a 6900 µF capacitor bank to the utility line to improve the power factor.

Next, we add a load current that draws a fifth-harmonic current of 50 A (Figure 4.19). The fifth-harmonic current results in significant load voltage distortion.

Next, we'll add a load that has a seventh-harmonic amplitude of 30.0 A, an 11th-harmonic (15.0 A), a 13th-harmonic (7.0 A), and 17th-harmonic (3.0 A). The resultant waveform (Figure 4.20) has 25 percent THD current distortion and 28.5 percent load voltage distortion. We'll see in a later chapter how one might add line filters to improve the load voltage.

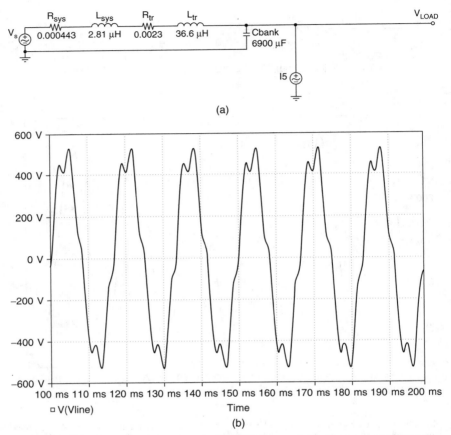

(a)

(b)

Figure 4.19 The circuit for Example 4.5 with fifth-harmonic load current added. (a) The circuit. (b) The load voltage.

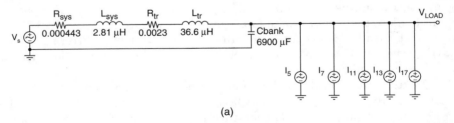

(a)

Figure 4.20 The circuit for Example 4.5, with 5th, 7th, 11th, 13th, and 17th harmonics. (a) The circuit. (b) The load voltage.

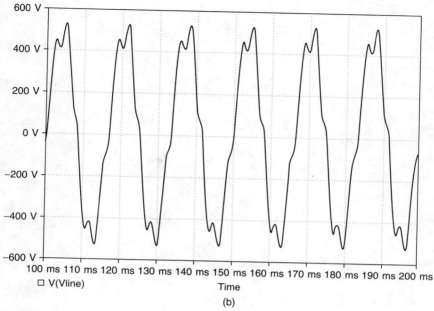

Figure 4.20   (*Continued*)

Summary

In this chapter, we developed harmonic analysis tools that will allow us to analyze waveforms and thus determine harmonic content. Harmonics cause many detrimental effects in equipment. In later chapters, we'll discuss methods of reducing harmonics.

References

[4.1] M. Holland, "Fundamentals on Harmonics," *Proceedings of the 1999 Cement Industry Technical Conference, IAS/PCA*, April 11–15, 1999, pp. 55–67.
[4.2] J. Fourier, *The Analytical Theory of Heat* (translated by A. Freeman). New York: Dover Publications, Inc., 1955. First published as *Théorie Analytique de la Chaleur*, by Firmin Didot, Paris, 1822.
[4.3] Richard Lee Ozenbaugh, *EMI Filter Design*, Marcel Dekker, 1996.
[4.4] M. Mardiguian, M., *EMI Troubleshooting Techniques*, McGraw-Hill, 1999.
[4.5] National Semiconductor, application note AN-990.
[4.6] R. W. Erickson and D. Maksimovic, *Fundamentals of Power Electronics*, 2nd ed, Springer, 2001.

Harmonic Current Sources

In this chapter, we shall discuss the circuits and magnetics that create harmonics (multiples of the line frequency) and/or interharmonics (other high-frequency components). Harmonics are generated by rectifiers, line-frequency converters, and nonlinear magnetics. Interharmonics are created by high-frequency switching power supplies. For most of what we shall do in this chapter, the fundamental frequency used will be 60 Hz.

Background

A typical setup that shows how harmonic currents can affect power quality is shown in Figure 5.1a. An AC voltage source is displayed, with its associated line reactance, X_s, and resistance, R_s. This AC source can be single-phase or three-phase. The line inductance depends on the length of the line and the geometry of the conductors. The line resistance, on the other hand, depends on the length of the wire and the wire gauge used. The AC source voltage then supplies a nonlinear load that draws harmonic current. Typically, this harmonic source is a rectifier or other converter.

In Figure 5.1b, we see the single-phase equivalent circuit. Note that the voltage labeled V_{pcc} (for *voltage at the point of common coupling* or *PCC*) has harmonic components due to the harmonic current I_h drawn by the load running through the line impedance. If this voltage at the PCC feeds additional equipment other than the harmonic generating circuit, the resulting voltage distortion can disrupt operation of the equipment if the harmonic distortion is too high. Harmonic limits are discussed in great detail in IEEE Std. 519.

In the following, we'll discuss numerous pieces of equipment that generate harmonic currents.

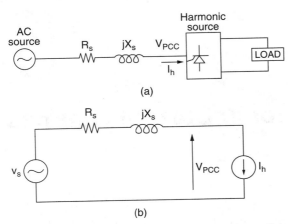

Figure 5.1   Illustration of a harmonic current source affecting power-quality. (a) A single-line diagram of an AC source
with source impedance $R_s + jX_s$ and a load that draws harmonic current I_h. (b) A single-phase equivalent circuit.

Single-Phase Rectifiers

Rectifiers are used in all sorts of power system and power electronic subsystems to convert AC power to DC power. In low-power applications
using single-phase power, rectifiers are used as the front-end of switching power supplies and small motor drives.

A single-phase, full-wave rectifier with current source load is shown
in Figure 5.2a. This circuit is an idealized model of systems where the
load draws approximately constant current.[1] Figure 5.2b shows the line
current. In this simplified model, the line current is a square wave. In
Figure 5.2c, we see the spectrum of the line current, where the first harmonic has amplitude 1.0, the third harmonic has amplitude 1/3, the fifth
harmonic has amplitude 1/5, and so on. Note that the line current drawn
by this rectifier circuit is very harmonic-rich, with a THD of 48.3 percent.

Another rectifier is the full-wave rectifier with capacitive filter
(Figure 5.3). In this type of circuit, the diodes are only on for a fraction of a 60 Hz cycle, and the capacitor charges near the peaks of the
input sine wave voltage. Therefore, the line current contains significant
harmonic distortion, as we'll see in the next example.

Example 5.1: Line-current harmonics.   In this example, we'll examine the
line-current harmonics drawn by the single-phase full-wave rectifier with

---

[1] For instance, this current source models a series LR load where L/R >> 1/60.

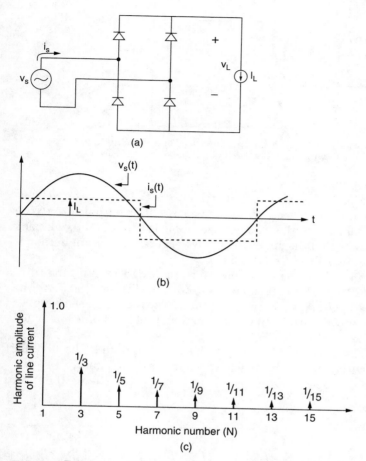

Figure 5.2 Full-wave rectifier with current source load. (a) The circuit. (b) Waveforms, showing line-voltage and line current. Line current is a square wave. (c) The spectrum of the line current, showing the characteristic 1/N rolloff in spectral component amplitude with harmonic number.

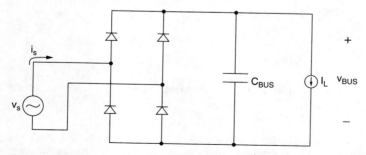

Figure 5.3 A full-wave rectifier with current source load, I_L, and capacitive filter, C_{BUS}.

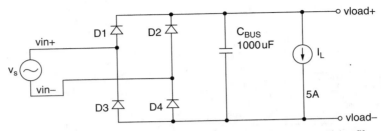

Figure 5.4 A full-wave rectifier with current source load and capacitive filter showing PSPICE circuit with bus capacitor 1000 μF and a load that draws a DC current of 5 A.

current source load and capacitive filter (Figure 5.4). In this case, we have a 120-V AC source, a DC load of 5 A, and a capacitive output filter. With a 1000 μF bus capacitor, the load voltage ripple is about 25-V peak-peak (Figure 5.5a). Note that the line current waveform has significant harmonic distortion (Figure 5.5b and Figure 5.5c).

Next, we can reduce the load voltage ripple by increasing the filter capacitor from 1000 μF to 10,000 μF. As shown in Figure 5.6a, note the peak-peak load voltage ripple is reduced, but at the expense of higher peak line current (Figure 5.6b) and higher levels of harmonic distortion in the line current (Figure 5.6c).

In Table 5.1, we see the spectra in both cases, tabulated. The THD for the rectifier with a 1000 μF capacitive filter is approximately 140 percent, and the peak-peak output ripple is 25 V. By increasing the filter

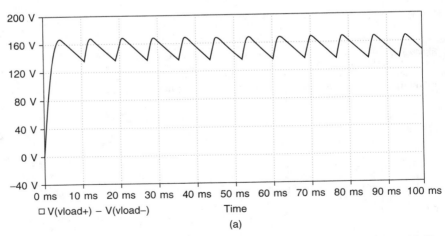

Figure 5.5 A full-wave rectifier with current source load and capacitive filter. (a) The waveform of the load voltage, which has a ripple due to the load current and the finite value of the filter capacitor. (b) The line current, which is very "spikey." (c) The spectrum of the line current.

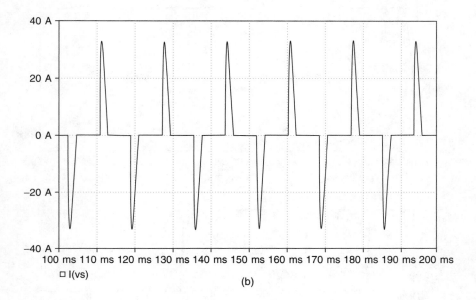

(b)

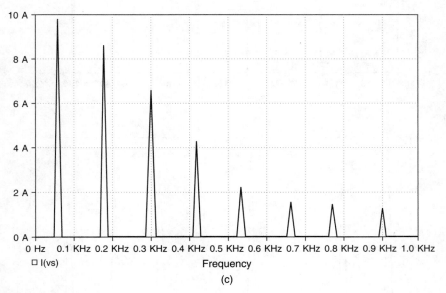

(c)

Figure 5.5 (*Continued*)

capacitor to 10,000 µF, we reduce the output ripple to approximately 4 V, but the line current is much more spikey and the THD of the line current is approximately 265 percent. This result illustrates one trade-off with this type of rectifier.

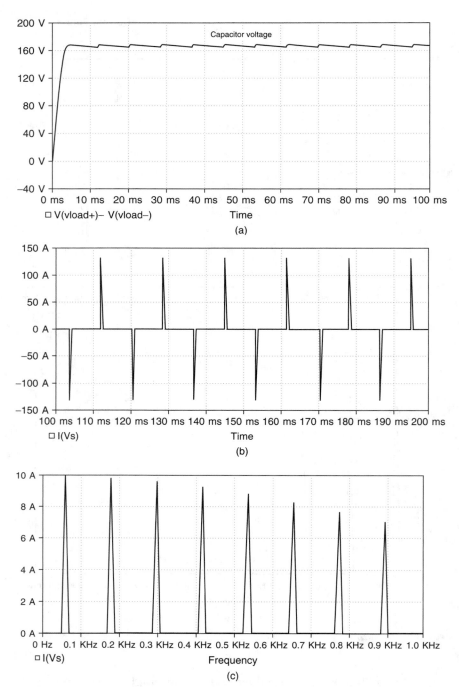

Figure 5.6 A full-wave rectifier with current source load and increased capacitive filter to 10,000 μF. (a) The waveform of the load voltage, which has a reduced ripple compared to the 1000 μF case. (b) The line current, which is very "spikey." (c) The spectrum of the line current.

TABLE 5.1 Results of the PSPICE Simulation Showing the Fourier Spectrum of Line Current for Example 5.1

Harmonic number	Harmonic current with 1000 μF filter capacitor	Harmonic current with 10,000 μF filter capacitor
1	10.0	10.0
3	9.4	10.0
5	6.9	9.8
7	4.8	9.5
9	3.0	9.0
11	2.2	8.25
13	2.1	7.75
15	2.0	7.0
17	1.65	6.5
19	1.35	6.3
21	1.25	5.0
23	1.2	4.5
25	1.1	3.8
27	1.0	3.3

Three-Phase Rectifiers

The six-pulse rectifier

A typical application using a three-phase, six-pulse rectifier is an adjustable speed drive (Figure 5.7). Three-phase power (labeled phases a, b, and c) is full-wave rectified by the six-pulse rectifier. The rectified voltage is filtered by the high-voltage bus capacitor, C_{bus}, generating a DC voltage, which is used by the subsequent inverter. The three-phase inverter generates the three-phase currents necessary to drive the motor.

A six-pulse rectifier is shown in Figure 5.8a. Assuming that the load approximates a current source (with very large load inductance), the line current drawn from the rectifier shows a THD of 31 percent, and an absence of a third-harmonic and all triplen harmonics (Figure 5.8b). We can show that the harmonic amplitudes of the phase currents of the ideal six-pulse rectifier with current source load, I_L, are [5.1]:

$$\frac{4}{N\pi} I_L \sin\left(\frac{N\pi}{2}\right) \sin\left(\frac{N\pi}{3}\right)$$

The spectrum of the AC line currents is shown in Figure 5.8c.

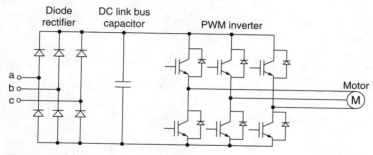

Figure 5.7 Adjustable speed drive for a three-phase induction motor.

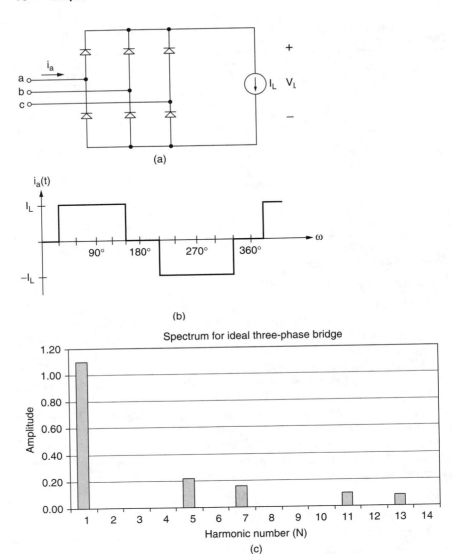

Figure 5.8 A six-pulse rectifier. (a) The circuit. (b) The line current for L/R >> period of line frequency. (c) The spectrum of the line current, assuming $I_L = 1$.

The twelve-pulse rectifier

The twelve-pulse rectifier (Figure 5.9a) is comprised of two six-pulse rectifiers fed from separate transformers. One six-pulse is fed from a Y/Y transformer, and the other is fed from a Δ/Y transformer. The two rectifier voltages are phase shifted 30° from one another. The resultant phase current waveform (Figure 5.9b) more closely mimics an ideal sine wave

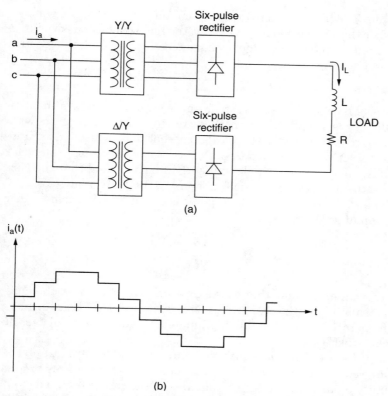

Figure 5.9 A twelve-pulse rectifier. (a) The circuit. (b) The line current for phase-a for L/R >> period of line frequency.

than in the six-pulse case. This is because the 12-pulse topology elimi-nates the 5th, 7th, 17th, and 19th harmonics, leaving the 11th, 13th, 23rd, and 25th harmonics.

High-Frequency Fluorescent Ballasts

High-frequency fluorescent ballasts (Figure 5.10) are a rich source of high-frequency voltage and current harmonics. The line waveform is rectified by a power-factor corrected boost converter. The rectified DC voltage powers a high-frequency inverter that generates the high voltage needed

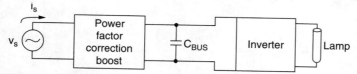

Figure 5.10 High-frequency switching fluorescent ballast with a power factor correction front end.

by the lamp. The power factor correction circuit switches at a frequency much higher than line frequency. Usually, the line side has passive filtering to help reduce the high-frequency harmonics drawn from the AC line.

Transformers

One source of harmonics using transformers with iron cores is the nonlinear B/H curve associated with these devices. Figure 5.11 shows a representative B/H curve for electrical steel that may be used in a transformer core. Not shown here, purposely, is the hysteresis loop, since for simplicity we'll consider a single-valued B/H curve. Note that the core material saturates at a magnetic flux density B ~ 2 Tesla or so.

The slope of the B/H curve at any point is the magnetic permeability of the core material, or

$$\mu_c = \frac{dB}{dH}$$

From this equation, note that as the core material saturates, the magnetic permeability decreases. For economic reasons, power transformers are often operated at a relatively high magnetic flux density, above the B/H curve "knee."

Shown in Figure 5.12a is a magnetic circuit without an airgap. The magnetic circuit is comprised of N turns, the core material has magnetic permeability μ_c, and the cross-sectional area of the core is A with a characteristic length, l_c, which is the mean path length of the flux lines inside

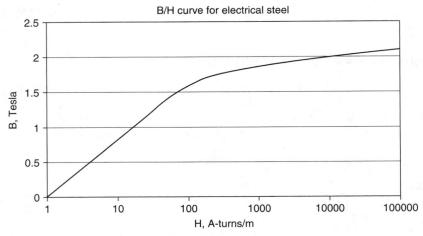

Figure 5.11   A representative nonlinear B/H curve.

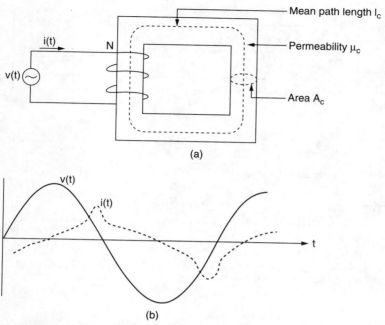

Figure 5.12   An inductor that draws nonlinear current. (a) The circuit. The inductor has N turns; a mean path length, l_c; core permeability, μ_c; and a core cross-sectional area, A_c. (b) The waveforms of inductor voltage, $v(t)$, and the inductor current, $i(t)$.

the core. From magnetic circuit analysis, we can find the inductance of this structure as:

$$L = \frac{\mu_c A N^2}{l_c}$$

We'll now assume that we excite this circuit with sufficient volt-seconds so the core material begins saturation. Equivalently, saturation means the permeability of the core decreases with a corresponding decrease in the inductance at high current levels. This nonlinear property of the magnetic core requires that the current has harmonic distortion, as shown in Figure 5.12b.

Other Systems that Draw Harmonic Currents

High-frequency switch-mode power supplies are covered extensively in Chapter 7. Other sources of harmonic currents are adjustable speed drives (ASDs), motors, and arc furnaces. Figure 5.13 illustrates the typical line current and spectrum for an adjustable speed drive.

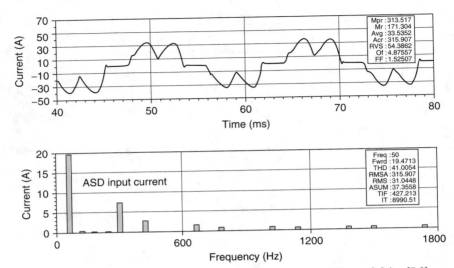

Figure 5.13 The input line current and spectrum for an adjustable speed drive [5.2]. [© 1995, IEEE, reprinted with permission]

Summary

In this chapter, we've covered equipment that generates harmonics, such as rectifiers, transformers, and switching power supplies. Harmonics can cause many detrimental effects, including resonances with power factor correction capacitors, the heating of neutral conductors, the false tripping of relaying equipment, and the heating of capacitors. In the next chapter, we'll discuss power harmonic filters.

References

[5.1] R. W. Erickson and D. Maksimovic, *Fundamentals of Power Electronics*, second edition, Springer, 2001.
[5.2] IEEE, "IEEE Recommended Practice for Monitoring Electric Power Quality," IEEE Std. 1159-1995.

Power Harmonic Filters

In this chapter, we will discuss methods of reducing harmonic distortion in line voltages and currents through the use of filters. Filters can be implemented with either passive components (capacitors and magnetics) or active filters. Here, we will examine filtering techniques applied to harmonics of 60 Hz, and to high-frequency "interharmonics" as well. The eventual goal of the use of such filters is to reduce harmonic distortion to within IEEE Std. 519 limits.

Introduction

Industrial and commercial power systems usually incorporate power capacitors to improve the power factor and provide reactive power for voltage support [6.1]. When the system includes sources of harmonic current, such as power electronic converters or adjustable speed drives (ASDs), the capacitors may be used in power harmonic filters to minimize the harmonic voltage applied to the system load at the point of common coupling (PCC).

The current harmonics produced by power converters, usually polyphase rectifiers, can be reduced in one of three ways: (1) series reactors in the input line; (2) the use of a 12-pulse, or higher, connection of the rectifier bridges, and (3) use of pulse-width modulation of the line current. When these measures do not reduce the current harmonics to an acceptable level, power harmonic filters can be introduced to obtain further reduction.

The current harmonics, of themselves, are seldom the problem, such as when the third harmonic produces overheating in the three-phase feeder neutral conductor. The problem occurs when a higher-order current harmonic is resonant with the capacitors and system reactance to produce excessive voltages at the point of common coupling (PCC).

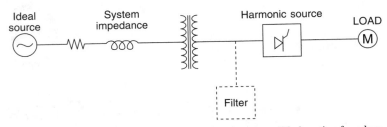

Figure 6.1   A typical distribution system showing a possible location for a harmonic filter.

A model of a distribution system powering a nonlinear load is shown in Figure 6.1. The utility is modeled as a source with impedance consisting of line resistance and line inductance. The source is then stepped down with a transformer. The resulting voltage (typically 480 V line-to-line in three-phase systems) drives nonlinear loads such as motors and other equipment.

Some of the goals of IEEE Standard 519 are that the utility presents good quality voltage to the load, and that the load doesn't draw overly high harmonic currents from the utility. We'll see several methods in the upcoming sections by which the effects of harmonic currents can be mitigated. Shown in Figure 6.1 is a typical location where a harmonic filter can be added.

A Typical Power System

A simplified version of a power system is shown in Figure 6.2 [6.2]. The power-factor correction capacitor has been converted to a series-tuned passive filter. In the nonlinear load, for example, an ASD requires a fundamental frequency current for operation, but can be represented as the source of harmonic currents into the system.

The paths of the harmonic currents produced by the "nonlinear load" of Figure 6.2 are shown in the diagram of Figure 6.3. The harmonic currents and voltages are described as follows:

- I_h: One harmonic component of the converter current—for example, fifth harmonic

- I_{hc}: The power-factor capacitor current before capacitor C became part of filter at location 3

- I_{hf}: The filter harmonic current

- I_{hc}: The corrected harmonic component injected into point of common coupling (PCC)

- I_{hl}: The equivalent passive motor load current

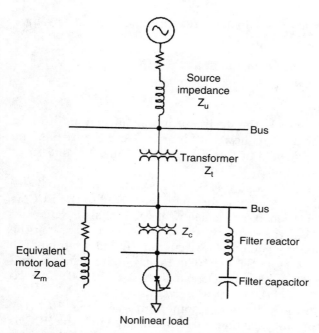

Figure 6.2 A power system equivalent circuit [6.2]. A motor load, nonlinear load, and a filter reactor are part of a power factor correction circuit.
[© 2004, IEEE, reprinted with permission]

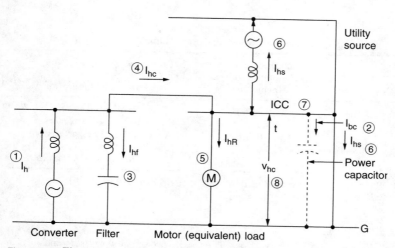

Figure 6.3 Electrical model of the system, including nonlinear load.

- I_{hs}: The harmonic current component returning to the utility source
- PCC: Point of common coupling
- V_h: The harmonic voltage component at the PCC

IEEE Std. 519-1992

IEEE Std. 519 [6.3] controls the design of power-harmonic filters in electrical systems such as that shown in Figure 6.2 and Figure 6.3. The standard does this by means of the Tables 10.1 and 10.3, shown in Figure 6.4 and Figure 6.5.

Table 10.1 (shown in Figure 6.4) sets the maximum individual frequency voltage harmonics (percent) for loads connected to the PCC as a function of the size of the load. The measure of size is short-circuit ratio (SCR), defined as I_{SC}/I_L. I_{SC} is the maximum short-circuit current at the PCC, and I_L is the maximum demand load current (fundamental) at the PCC. A very large load has an SCR of 10 and maximum harmonic voltages of 2.5 to 3.0 percent. A very small load has an SCR of 1000 and maximum harmonic voltages of 0.05 to 0.10 percent.

In order to achieve the voltage harmonic limits of Table 10.1, the standard sets limits on the harmonic currents injected into the PCC as a function of the size of the load in Table 10.3 of Figure 6.5. The harmonic current is shown in Figure 6.3 as I_{hc}. The current is the resultant of the converter current I_h and the filter current I_{hf}. For example, for a small load, $I_{SC}/I_L > 1000$, the current harmonics less than the 11th must be less than 15.0 percent of I_L. In addition, the TDD (total demand distortion[1]) must be less than 12.0 percent.

Table 10.1
Basis for Harmonic Current Limits

SCR at PCC	Maximum Individual Frequency Voltage Harmonic (%)	Related Assumption
10	2.5–3.0%	Dedicated system
20	2.0–2.5%	1–2 large customers
50	1.0–1.5%	A few relatively large customers
100	0.5–1.0%	5–20 medium size customers
1000	0.05–0.10%	Many small customers

Figure 6.4   Basis for harmonic current limits, from IEEE Std. 519 [6.3].
[© 1992, IEEE, reprinted with permission]

---

[1] TDD is defined in IEEE-519, p. 11, as "The total root-sum-square harmonic current distortion, in percent of the maximum demand load current..."

Table 10.3
Current Distortion Limits for General Distribution Systems
(120 V Through 69 000 V)

	Maximum Harmonic Current Distortion in Percent of I_L					
	Individual Harmonic Order (Odd Harmonics)					
I_{sc}/I_L	<11	$11 \leq h < 17$	$17 \leq h < 23$	$23 \leq h < 35$	$35 \leq h$	TDD
<20*	4.0	2.0	1.5	0.6	0.3	5.0
20 < 50	7.0	3.5	2.5	1.0	0.5	8.0
50 < 100	10.0	4.5	4.0	1.5	0.7	12.0
100 < 1000	12.0	5.5	5.0	2.0	1.0	15.0
>1000	15.0	7.0	6.0	2.5	1.4	20.0

Even harmonics are limited to 25% of the odd harmonic limits above.

Current distortions that result in a dc offset, e.g., half-wave converters, are not allowed.

*All power generation equipment is limited to these values of current distortion, regardless of actual I_{sc}/I_L.

where
I_{sc} = maximum short-circuit current at PCC.
I_L = maximum demand load current (fundamental frequency component) at PCC.

Figure 6.5 Current distortion limits, from IEEE Std. 519 [6.3].
[© 1992, IEEE, reprinted with permission]

The designer of the system, whose goal it is to comply with IEEE Std. 519 with a given load connected to the PCC must either control the converter harmonic currents or design a suitable filter to comply with Table 10.3. Next, we'll discuss various filtering methods that may be used to comply with this IEEE standard.

Line reactor

One of the simplest harmonic filters is the line reactor[2] shown as the three-legged inductor in Figure 6.6. This magnetic component is often used in the line in series with motor controllers and other converters that draw significant harmonic current. The reactor presents high impedance to high frequency currents while passing the fundamental.

The theoretical waveform of the line current of the six-pulse converter (rectifier) is shown in Figure 6.7a. This first figure assumes no line inductance. When we add a line reactor, the inductance of the reactor causes the converter to exhibit a significant commutation time. The

---

[2] The line reactor is basically a series inductor. Remember that the magnitude of the impedance of an inductor is $2\pi fL$.

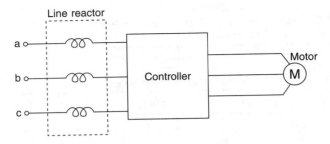

Figure 6.6 The line reactor in a motor controller circuit.

result is the trapezoidal waveform of Figure 6.7b, where the commutation time is the interval μ. The reactor does not reduce the 5th and 7th harmonics significantly, but does reduce the 11th harmonic and higher—for example, the 11th is reduced from 9.1 to 6.5 percent and the 23rd from 4.3 to 0.5 percent [6.2]. The reactor will also reduce the magnitude of the short-circuit current within the converter.

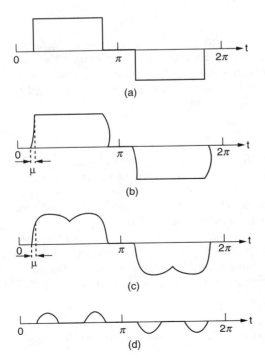

Figure 6.7 Waveforms of a six-pulse converter [6.2].
[© 2004, IEEE, reprinted with permission]

Figure 6.8   Photograph of a 0.047 mH three-phase line reactor.

In Figure 6.8, we see a photograph of a line reactor rated at 0.047 millihenries, or 0.0178 Ω at 60 Hz. For line reactor applications, the reactor is usually rated in the percent voltage drop at the rated load current. For a rated load of 500 A on a 480/277-V system, the reactor would be rated at (500 A × 0.0178 Ω) × 100/277 V = 3.2 percent reactance.

Shown in Figure 6.9 [6.4] are line current waveforms to an adjustable speed drive. The diagram illustrates the typical reduction in harmonics and total harmonic distortion (THD) that can be accomplished through the use of line reactors. Without the reactor, we see a THD of 80.6 percent. With the addition of a 3-percent reactor,[3] we see a significant reduction in the "spikiness" of the current waveform, and a corresponding reduction in harmonic content and THD to 37.7 percent.

Shunt passive filter

The installation of the shunt passive filter is the most common method for controlling harmonic currents and achieving compliance with IEEE Std 519. The filter is usually placed as shown in Figure 6.3 to divert a selected portion of the harmonic currents produced by the

---

[3] Again, the value of this inductor is specified in the per-unit system.

nonlinear load. The capacitors of the filter also provide reactive power at the fundamental frequency (60 Hz) for power factor correction. The filter is usually made up of one or more sections, as shown in Figure 6.10 [6.5]. The single-tuned RLC filter for each harmonic frequency is the most common.

The impedance Z of the single-tuned section shown in Figure 6.11a is given by:

$$Z = R + j(X_L - X_c)$$

where:

$$X_L = \omega L$$

$$X_C = \frac{1}{\omega C}$$

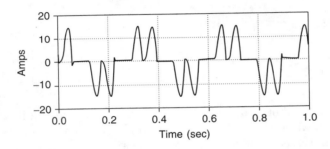

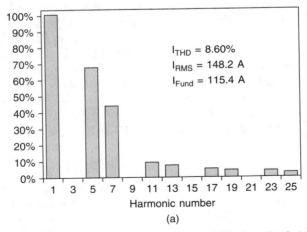

(a)

Figure 6.9 Input waveforms to a 100-hp ASD, from [6.4]. (a) Without an input line reactor, the line current THD = 80.6 percent; (b) with a 3-percent line reactor, the line current THD = 37.7 percent.
[© 1999, IEEE, reprinted with permission]

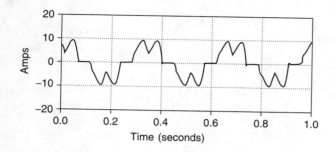

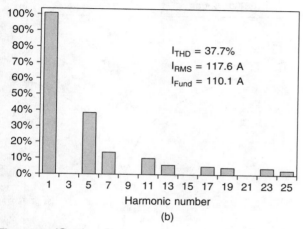

(b)

Figure 6.9 (*Continued*)

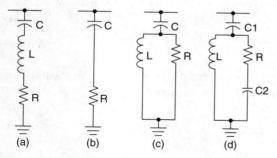

(a) Single-tuned filter
(b) First order high-pass filter
(c) Second order high-pass filter
(d) Third order high-pass filter

Figure 6.10   Shunt filters [6.5].
[© 1985, IEEE, reprinted with permission]

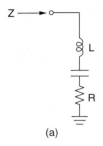

(a)

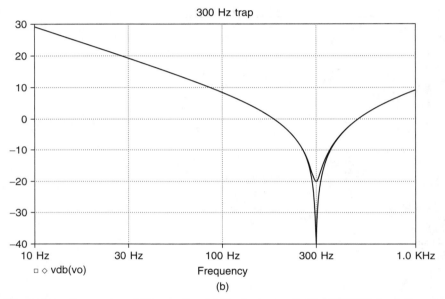

(b)

Figure 6.11   A series-tuned filter section. (a) The basic circuit. (b) A PSPICE simulation of input impedance Z of a section tuned to the fifth-line harmonic (300Hz) with $L = 500$ μH, $C = 563$ μF, and $R = 0.01$ Ω and 0.1 Ω.

The resistance R is due to the winding loss and the core loss of the inductor. The quality factor, or Q of an inductor, is given by:

$$Q = \frac{\omega L}{R} = \frac{X_L}{R}$$

Typical values of Q for filter inductors are 30 to 50 at 60 Hz.

The series resonant circuit has a dip in its series impedance at resonance at the frequency where the inductive impedance and capacitive impedance exactly cancel each other out. The single-tuned section is at resonance for a frequency ω_r, where $X_L = X_C$, or $\omega_r = 1/\sqrt{LC}$. At resonance, the series impedance reduces to $Z = R$. For other than resonance, the magnitude of Z is given by:

$$|Z| = \frac{\sqrt{(RC\omega)^2 + (1 - LC\omega^2)^2}}{\omega C}$$

The impedance of a single-tuned section tuned to 300 Hz, or $\omega_r = 1884$ rad/s, is shown in Figure 6.11b. Two values of R are shown, $R = 0.01\ \Omega$ and $0.1\ \Omega$, which becomes the impedance at resonance. The impedances of the L and C at resonance are selected for the example as $0.94\ \Omega$.

Example 6.1: Series resonant filter.   A series resonant filter used on an AC line is shown in Figure 6.12a. We see the series resonant circuit L_f, C_f,

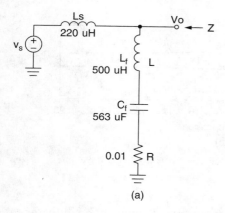

(a)

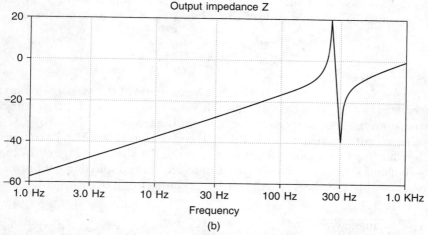

(b)

Figure 6.12   A series-tuned filter section example. (a) The basic circuit. (b) The magnitude of the output impedance Z at the PCC.

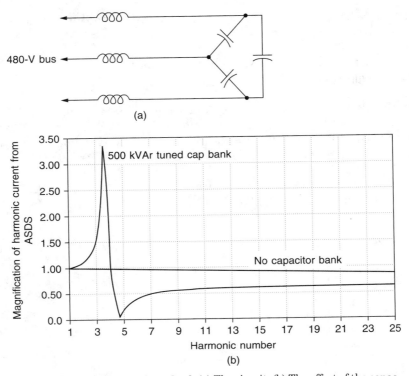

(a)

(b)

Figure 6.13   A three-phase filter [6.4]. (a) The circuit. (b) The effect of the capacitor bank on the voltage at the PCC.
[© 1999, IEEE, reprinted with permission]

and R_1 that has been added to the line. Inductor L_s models the inductance
of the source. The magnitude of the impedance Z at the output node is

$$|Z| \approx \frac{(\omega L_s)\sqrt{(1 - (L_f C_f \omega^2))^2}}{\sqrt{(1 - ((L_f + L_s)C_f \omega^2))^2}}$$

We see that the transfer function has a minimum at the series resonant
frequency of the filter components. There is also a peak at a frequency
lower than the series resonant frequency at the frequency where $(L_s + L_f)$
resonates with the capacitor. A PSPICE simulation of the fifth-harmonic
filter showing the magnitude of the impedance Z is shown in Figure 6.12b.

A harmonic filter designed to attenuate fifth-harmonic components for
an adjustable speed drive[4] application is shown in Figure 6.13a [6.4]. In

---

[4] In the power world, capacitors are often specified not by the capacitance value, but by
the VAr rating. VAr stands for volt-amperes reactive. The VAr rating of a capacitor is found
using $VA_r = (V)(A) = (V)(V/\omega C) = (V^2/\omega C)$.

Figure 6.13b, we see the characteristic peak and valley of this type of harmonic filter. Harmonic filters have also been used to reduce harmonic interference with telephone systems.

Multisection filters

We can use multiple harmonic filter sections (Figure 6.14) to reduce the effects of higher-order harmonics generated by nonlinear loads connected to the PCC. The filter is designed to attenuate higher-order harmonics such as the 5th, 7th, and 11th that are generated by the nonlinear load. Generally, the filter components are tuned a few percent below the harmonic frequency [6.2] to account for component variations, temperature variations, component aging, and system changes.

Figure 6.15a shows a filter designed to attenuate the 5th, 7th, and 11th harmonics. In this design example, each filter section is tuned 4 percent below the filtered harmonic. The series resonant frequencies of the three series resonant circuits are

$$f_1 = \frac{1}{2\pi \sqrt{L_1 C_1}} = \frac{1}{2\pi \sqrt{(500 \times 10^{-6})(611 \times 10^{-6})}} = 288\,\text{Hz}$$

$$f_2 = \frac{1}{2\pi \sqrt{L_2 C_2}} = \frac{1}{2\pi \sqrt{(500 \times 10^{-6})(312 \times 10^{-6})}} = 403\,\text{Hz}$$

$$f_3 = \frac{1}{2\pi \sqrt{L_3 C_3}} = \frac{1}{2\pi \sqrt{(500 \times 10^{-6})(126 \times 10^{-6})}} = 634\,\text{Hz}$$

These are the frequencies at which we expect significant attenuation, as evidenced in the PSPICE plot of Figure 6.15b. We also see peaking at frequencies below the three series-resonant frequencies. This is characteristic of harmonic single-tuned filters and we need to be

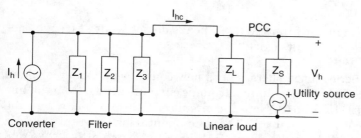

Figure 6.14 A multiple section filter. The harmonic generating load is modeled as a current source of value I_h.

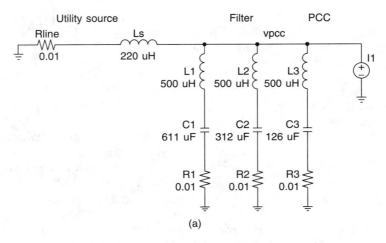

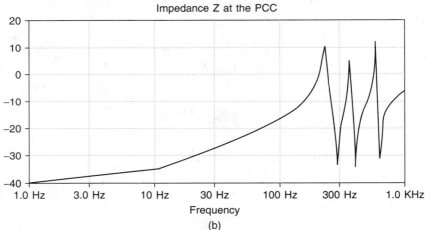

Figure 6.15 A harmonic multisection filter. This filter provides mimima at the 5th, 7th, and 11th harmonics. The line is modeled as an ideal voltage source, with a line inductance of 220 ΩH and a line resistance of 0.01 Ω. (a) The circuit. (b) A PSPICE simulation showing the impedance at the PCC.

mindful to design our filters so these peaks do not occur at frequencies where harmonic currents exist.

Example 6.2: Series-tuned filters.   Let's now consider an example where we'll design a filter to attenuate harmonic currents drawn from the line to comply with IEEE-519. Note the circuit of Figure 6.16a, where the source is 277 V, line-to-neutral. The fundamental load current at 60 Hz is $I_L = 100$ A. This load also draws fifth-harmonic current $I_5 = 20$ A and seventh-harmonic current $I_7 = 15$ A.

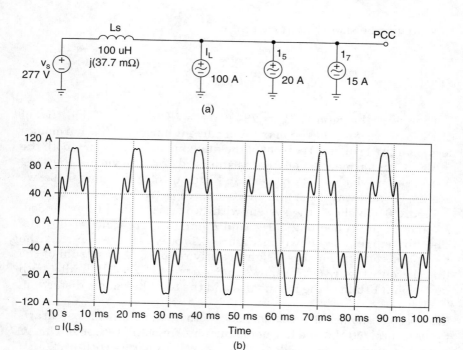

(a)

(b)

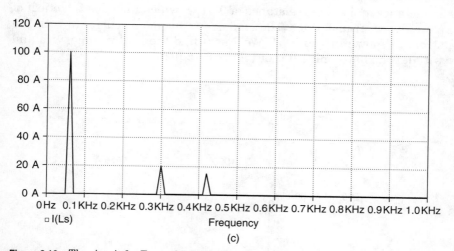

(c)

Figure 6.16 The circuit for Example 6.2. (a) The circuit. (b) The line current. (c) The spectrum of the line current.

First, we need to determine what the IEEE-519 limits are for line harmonic currents, using Table 10.3 (Figure 6.5). Note that the harmonic current limits depend on the ratio I_{SC}/I_L, where I_{SC} is the short-circuit current at the PCC, and $I_L = 100\,\text{A}$ is the fundamental load current. For this circuit, the short circuit current is

$$I_{SC} = \frac{V_s}{X_{L_s}} = \frac{277}{(2)(\pi)(60)(10^{-4})} = 7348 \text{ A}$$

Therefore, the ratio $I_{SC}/I_L = 7348/100 = 73.48$. From Table 10.3 of IEEE-519, we see that the maximum current harmonic (for harmonics less than the 11th) is 10 percent of the fundamental. Therefore, both the 5th and 7th harmonics violate this standard. We also violate the TDD specification, which is <12 percent. In Figure 6.16b and Figure 6.16c, we see the current waveform and spectrum, respectively.

Figure 6.17a shows the system with two series-tuned filters added. L_{f1} and C_{f1} are tuned to 290 Hz, and L_{f2} and C_{f2} are tuned to 407 Hz. The line current is shown in Figure 6.17b, where we see that the harmonic content has been significantly reduced. A spectrum of the line current (Figure 6.17c) shows that we meet IEEE-519 limits, both for the amplitude of the harmonics and the total harmonic distortion.

Example 6.3: another filter design.   A hypothetical nonlinear load draws fundamental (60 Hz) and harmonic currents from an AC source. The AC source is modeled as an ideal 277-V source in series with a 220-μH inductance and a line resistance of 0.01 Ω, which includes the effect of a step-down transformer. The load draws harmonic currents with the strength shown in Table 6.1. We'll next find the load voltage with and without the harmonic filter of Figure 6.15 using PSPICE.

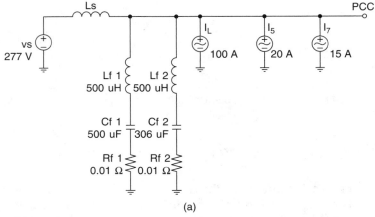

(a)

Figure 6.17   The circuit for Example 6.2, with series-tuned filters added (a) The circuit. (b) The line current. (c) The spectrum of the line current.

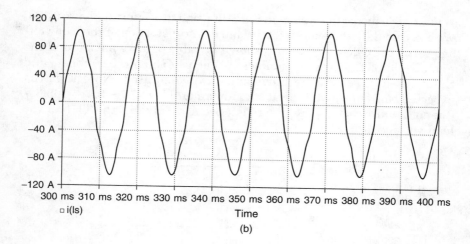

(b)

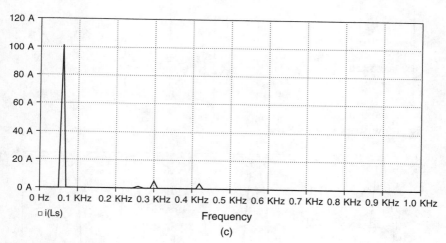

(c)

Figure 6.17 (*Continued*)

TABLE 6.1 Harmonic Currents for Example 6.1

Harmonic number	Value (Amps)
1	100
5	50
7	25
11	15

First, we'll model the system in PSPICE using the circuit of Figure 6.18a. We can calculate the expected total harmonic distortion of the load voltage using the calculated values shown in Table 6.2. We have found the magnitude of the line impedance at each harmonic frequency, and then the voltages at the harmonics are calculated. Note that the expected rms output voltage at the fundamental is $277 - 8.4 = 268.6$ V due to the drop across the line reactance at 60 Hz. So, the THD of the load voltage is

$$\text{THD} = \frac{\sqrt{20.7^2 + 14.5^2 + 13.7^2}}{268.6} = 10.7\%$$

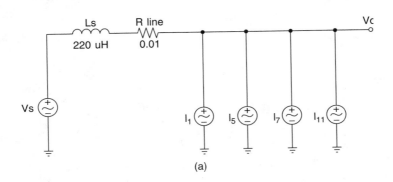

(a)

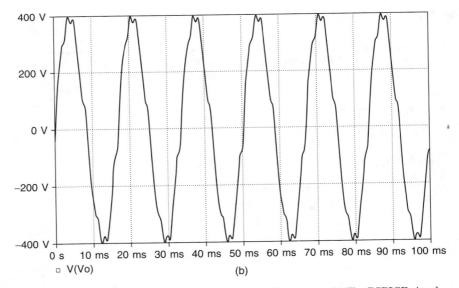

□ V(Vo)   (b)

Figure 6.18   The original system of Example 6.3. (a) The circuit. (b) The PSPICE simulation showing the load voltage. (c) The spectrum of the load voltage.

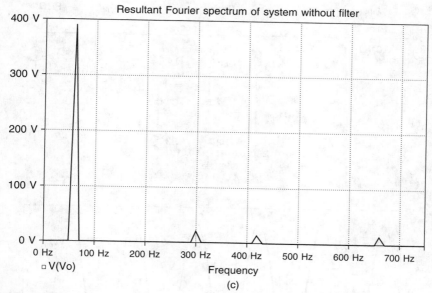

Figure 6.18   (*Continued*)

The PSPICE simulation result (Figure 6.18b) shows the result in the time domain. In the frequency domain, we see harmonic distortion, as expected, at the 5th, 7th, and 11th harmonics, as shown in the Fourier spectrum of Figure 6.18c.

We can reduce the harmonic distortion in the load voltage using the circuit of Figure 6.19a. The three series-resonant circuits are tuned 4 percent below the 5th, 7th, and 11th harmonics. In Figure 6.19b, we see an improvement in the waveform distortion. Looking at the spectrum (Figure 6.19c), note that the 5th, 7th, and 11th harmonic distortion has been largely eliminated and that the resultant THD of the load voltage in the filtered case is roughly 1 percent.

TABLE 6.2   Calculating THD for Example 6.3

Harmonic number	Magnitude of line impedance at harmonic frequency	Voltage drop at harmonic frequency
1	0.08 Ω	8.4
5	0.41 Ω	20.7
7	0.58 Ω	14.5
11	0.91 Ω	13.7

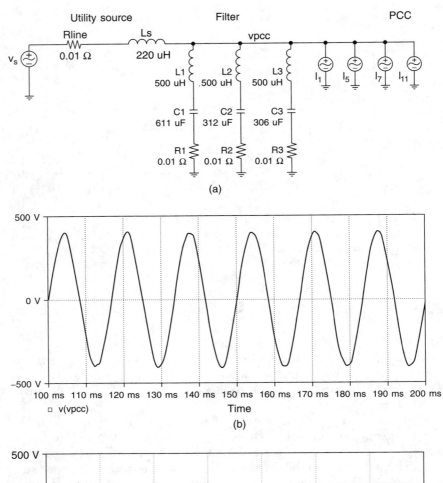

(a)

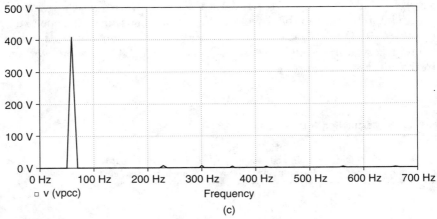

(b)

(c)

Figure 6.19 The system of Example 6.3 with single-tuned sections added. (a) The circuit. (b) The PSPICE simulation showing the load voltage. (c) The spectrum of the load voltage

Practical Considerations in the Use of Passive Filters

Passive power harmonic filter design requires a number of practical considerations. There is no unique solution to the design problem, so in each case a careful trade-off analysis must be performed. Practical considerations include the following:

- **Tuning:** The harmonic filter sections are tuned below the harmonic frequency to prevent the filter frequency from shifting upward if one or more capacitors fail and their fuses blow. Typical orders are 4.85 for the 5th harmonic; 6.7 for the 7th; and 10.6 for the 11th harmonic.

- **Protection:** Capacitors are protected by fuses in small groups to minimize the effect of fuse blowing. The whole filter can be divided into assemblies, each protected by a circuit breaker.

- **Switching:** Filters provide fundamental frequency reactive power (vars). Portions of the filter can be switched off at times of light load to limit overvoltage.

- **Tolerances:** Capacitors and inductors must be specified so that the combination of ratings (L and C) does not result in resonance at an undesired frequency. In other words, we do not want positive peaks in the filter impedance curves to correspond with harmonic frequencies.

- **Rating:** The current rating of the inductors and the voltage rating of the capacitors must include the fundamental and harmonic components.

- **Location:** Filters should be located electrically close to the nonlinear load that produces the harmonic currents.

- **Detuning:** A change in system impedance or component variations due to aging or temperature can result in some detuning of the harmonic filter.

Active harmonic filters

Active harmonic filters have been a growing area of research in recent years, due to improvements in switch technology, and also because of cost issues associated with filter components [6.6–6.8]. Especially at high power levels, the cost of magnetic and capacitive components can be high. High-frequency switching devices, including the metal-oxide semiconductor field-effect transistor (MOSFET) and insulated gate bipolar transistor (IGBT) have emerged in recent years with high current and voltage ratings. These devices switch on and off with fast switching speeds. Thus, high-frequency converters can be designed with good power delivery efficiency using them.

Passive filters can also introduce harmonic peaking (as we have demonstrated earlier), and the resonant networks can have high currents and voltages, resulting in high VAr ratings for the components. An alternative to passive filters are active filters, where power electronics components are used to actively inject harmonics to cancel harmonics in the line current. This method has been used in the past in lower power electronics applications [6.9].

The diagram of one type of active harmonic filter is shown in Figure 6.20. Note that the nonlinear load draws harmonic current, I_h, from the power source. The active compensator senses the harmonic current and injects a compensation current, I_c, which cancels the harmonic current. The net supply current, I_s, contains only the fundamental. The compensator switches at a very high frequency compared to the fundamental frequency—hence, the VA rating of the energy storage devices in the compensator can be minimized.

Purported advantages of active filters are [6.6]:

- **Superior filtering performance:** The active filters are generally under microprocessor control and hence can be tuned to a particular

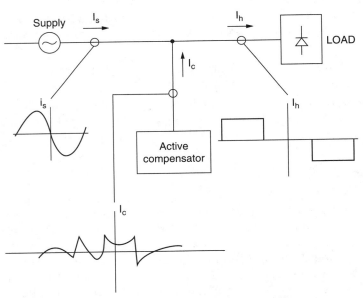

Figure 6.20 A typical active harmonic filter installation. The goal is to have the supply current be the only fundamental harmonic. The load draws harmonic current, I_h, and the active compensator injects a current, I_c, to cancel the harmonic content in the line current.

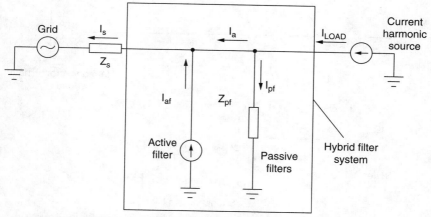

Figure 6.21 Hybrid filter including paralleled passive and active filters [6.10].
[© 2002, IEEE, reprinted with permission]

application. The filter can be tuned under microprocessor control if, for instance, the system impedance changes.

- **Smaller physical size:** The high switching speed of devices allows energy storage elements (capacitors and inductors) to be of smaller weight and volume.

- **Flexibility:** They are more flexible in application compared to passive filters.

Of course, the purported advantages of active filters must be weighed against extra design time and cost.

Hybrid harmonic filters

The application of harmonic filters including both passive and active elements has also been proposed [6.10] in so-called "hybrid" filters [6.11]–[6.13]. In this method, harmonic reduction and reactive power compensation is shared between a passive filter and a modest active filter. (Figure 6.21 and Figure 6.22). Typically, the active filter section is rated at a few percent of load kVA.

Summary

In this chapter, we have discussed power harmonic filters, both passive and active. Passive filters can be as simple as a line reactor, or as complicated as a multisection filter with individual sections tuned to resonant frequencies. Active filters afford design flexibility and smaller physical size. In all cases, filters are designed so that IEEE 519 limits are met.

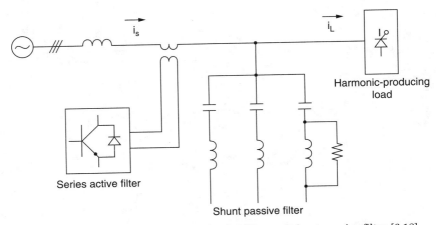

i_s

i_L

Harmonic-producing load

Series active filter

Shunt passive filter

Figure 6.22 The series connection of an active filter and shunt passive filter [6.12]. [© 1991, IEEE, reprinted with permission]

References

[6.1] J. K. Phipps, "A Transfer Function Approach to Harmonic Filter Design," *IEEE Industry Applications Magazine*, March/April 1997, pp. 68–82.

[6.2] J. C. Das, "Passive Filters—Potentialities and Limitations," *IEEE Transactions on Industry Applications*, vol. 40, no. 1, January/February, 2004, pp. 232–241.

[6.3] IEEE, "IEEE Recommended Practices and Requirements for Harmonic Control in Electrical Power Systems," IEEE Std. 519-1992, revision of IEEE Std. 519-1981.

[6.4] M. McGranaghan and D. Mueller, "Designing Harmonic Filters for Adjustable-Speed Drives to Comply with IEEE-519 Harmonic Limits," *IEEE Transactions on Industry Applications*, vol. 35, no. 2, March/April 1999, pp. 312–318.

[6.5] D. A. Gonzales and J. C. McCall, "Design of Filters to Reduce Harmonic Distortion in Industrial Power Systems," *Conference Record, IEEE-IAS-1985 Annual Meeting*, pp. 361–370.

[6.6] H. Akagi, "Active Harmonic Filters," *Proceedings of the IEEE*, vol. 93, no. 12, December 2005, pp. 2128–2141.

[6.7] Schneider Electric, Inc., "Proper Use of Active Harmonic Filters to Benefit Pulp and Paper Mills." Available from the Web at http://ecatalog.squared.com/pubs/Power%20Management/Power%20Quality%20Correction%20Equipment/Accusine%20PCS/5820DB0502.pdf.

[6.8] F. Z. Peng, "Application Issues of Active Power Filters," *IEEE Industry Applications Magazine*, September/October 1998, pp. 21–30.

[6.9] L. LaWhite and M. F. Schlecht, "Design of Active Ripple Filters in the 1-10 MHz Range," *IEEE Transactions on Power Electronics*, vol. 3, no. 3, July 1988, pp. 310–317.

[6.10] Z. Chen, F. Blaabjerg, and J. K. Pedersen, "A Study of Parallel Operations of Active and Passive Filters," *2002 Power Electronics Specialists Conference (PESC '2002)*, June 23–27, 2002, pp. 1021–1026.

[6.11] D. Rivas, L. Moran, J. Dixon, and J. Espinoza, "Improving Passive Filter Compensation Performance with Active Techniques," *IEEE Transactions on Industrial Electronics*, vol. 50, no. 1, February 2003, pp. 161–170.

[6.12] H. Fujita and H. Akagi, "A Practical Approach to Harmonic Compensation in Power Systems—Series Connection of Passive and Active Filters," *IEEE Transactions on Industry Applications*, vol. 27, no. 6, November/December 1991, pp. 1020–1025.

[6.13] S. Bhattacharya, P. Cheng, and D. Divan, "Hybrid Solutions for Improving Passive Filter Performance in High Power Applications," *IEEE Transactions on Industry Applications*, vol. 33, no. 3, May/June 1997, pp. 732–747.

7

Switch Mode Power Supplies

In this chapter, we shall examine high-frequency switching power supplies.[1] Such systems are used extensively in pulse-width modulated (PWM) inverters and in DC/DC converters. Typically, switching frequencies are in the few-kiloHertz range to upwards of a megahertz or higher. Due to the high switching speed of these power supplies, and also due to the fast rising and falling edges of voltages and currents, these converters create significant high-frequency harmonics.

Background

Switch mode power supplies are used extensively in consumer and industrial equipment such as personal computers and battery chargers. These systems require fast switching waveforms with high di/dt and dv/dt to ensure good power delivery efficiency.[2] These fast switching frequencies have a detrimental effect on electromagnetic compatibility (EMC) because of conducted electromagnetic interference (EMI) to the power line.[3] For instance, high-frequency switching components can be drawn from the AC power line if sufficient EMI filtering is not present.

---

[1] In this chapter, we shall use the terms "switch-mode power supply," "high frequency switching power supply," and "DC/DC converter" interchangeably. In each case, there is switching going on at frequencies much higher than the line frequency.

[2] In recent years, a trend has developed to require higher efficiency in offline power supplies for many consumer products. Agencies such as the California Energy Commission and the EPA (with its Energy Star program) are active in this arena.

[3] The mechanisms by which high-frequency currents from the switching action couple to the power line are beyond the scope of this work.

Conducted EMI can also result in radiated noise, since a primary source of radiated EMI is current in the AC power leads. In recent years, the United States and Europe have placed stringent requirements on the harmonic pollution injected into power lines. In this chapter, we will discuss high-frequency power supply issues and other technical challenges related to harmonic injection and EMI.

Offline Power Supplies

Offline power supplies are a class of supplies supplied by the AC line, and generally produce one or more DC outputs. The input AC source is rectified, and the resultant rectified voltage is chopped at a high frequency to produce the desired DC output(s). The load presented by the input rectifier is a large capacitor, so the line current is therefore highly discontinuous, as shown in the typical waveform of Figure 7.1a. A typical spectrum of the input current is shown in Figure 7.1b.

Numerous types of switching power supplies are used in offline applications, and we will discuss some representative examples. Figure 7.2 shows a flyback converter, which is often used in low-cost, low-power (less than a few hundred watts) applications, where isolation is needed between the line-voltage and the output voltage. The flyback provides

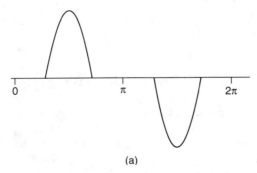

(a)

Table 4.3
Spectrum of Typical Switch Mode Power Supply

Harmonic	Magnitude	Harmonic	Magnitude
1	1.000	9	0.157
3	0.810	11	0.024
5	0.606	13	0.063
7	0.370	15	0.079

(b)

Figure 7.1   An offline power supply that draws its power from the line through a full-wave rectifier [7.1]. (a) A typical line-current waveform. (b) Typical spectrum of a switching power supply.
[© 1992, IEEE, reprinted with permission]

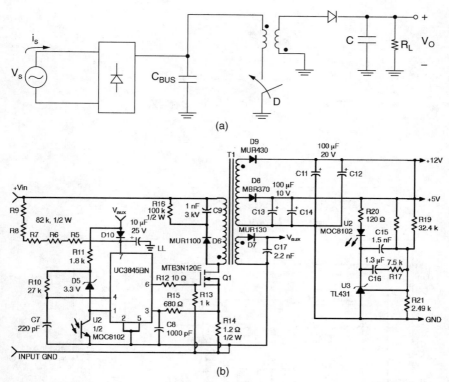

Figure 7.2   An offline flyback converter that draws its power from the line through a full-wave rectifier. (a) The switch is turned on and off at a frequency that is high compared to line frequency. The switch is ON for a fraction of a switching cycle of D percent. (b) The circuit, from On-Semiconductor Application Note AN-1327 [7.2], with the input rectifier not shown.
[© 2002, On Semiconductor, reprinted with permission]

a DC output voltage, and typical applications include battery chargers, high voltage power for displays, and other low-cost power supplies.

A full-wave rectifier at the front end of the flyback converter converts the AC input waveform to a DC waveform, which is filtered by the bus capacitor, C_{BUS}. The voltage across the bus capacitor is used to periodically energize the primary side of the isolation transformer. When the switch turns off, energy is dumped to the output capacitor. This process repeats itself over and over at the switching frequency, f_{sw}. This switching action generates frequency components at multiples of the switching frequency f_{sw} to be drawn from the line through the full-wave rectifier. The bus capacitor will have some series inductance, and at high frequencies some of the switch current can bypass the bus capacitor and enter the utility line.

Another type of DC/DC converter is the boost converter (Figure 7.3). The boost converter is being used more and more in consumer and industrial

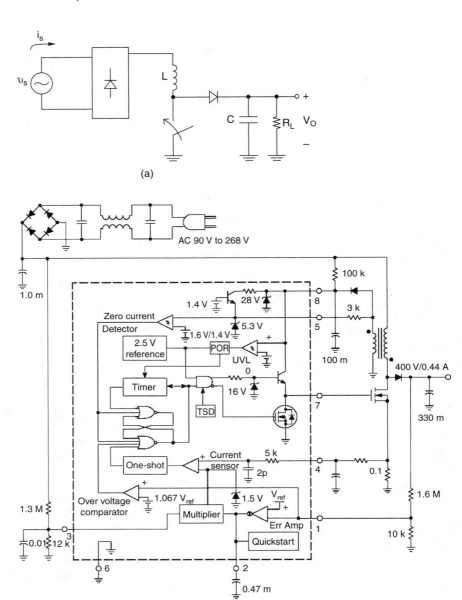

Figure 7.3   An offline boost converter. (a) A basic circuit. (b) Implementation using the On Semiconductor MC33232 Power Factor Controller [7.3].
[© 2005, On Semiconductor, reprinted with permission]

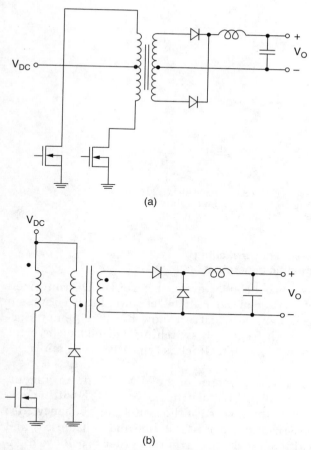

(a)

(b)

Figure 7.4   Other high-frequency DC/DC converters. (a) The push-pull converter. (b) The forward converter.

electronics since it can be configured to provide line-frequency power-factor correction. By properly controlling the high-frequency switch, the first harmonic of the line current is forced to be sinusoidal and in-phase with the line voltage, improving power factor to be near 100 percent.

These and other DC/DC converter topologies (including the buck converter, the buck-boost, the Cuk converter, the SEPIC converter, the full-bridge converter, the forward converter, and the push-pull converter) are commonly used. The push-pull and forward converters are shown in Figure 7.4. The push-pull and forward DC/DC converters are used in medium- to high-power applications.[4]

---

[4] In the world of DC/DC converters, "medium- and high-power" applications are generally in the multiple 100s of watts to kilowatts.

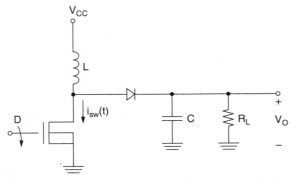

Figure 7.5 Boost DC/DC converter. The converter switches ON and OFF at a switching frequency f_{sw} and a duty cycle D. The duty cycle D is the fraction of the switching cycle that the MOSFET is ON.

DC/DC Converter high-frequency switching waveforms and interharmonic generation

Shown in Figure 7.5 is a boost converter, one type of DC/DC converter. Boost converters are often used in the front end of power factor correction circuits, so we'll use it here to illustrate how high-frequency harmonics[5] are created by high-frequency switching. The boost converter has a switching frequency $f_{sw} = 1/T$, which is typically in the multiple-kHz range

Shown in Figure 7.6a is the waveform of the MOSFET drain current, labeled $i_{sw}(t)$. Note that the MOSFET switches ON and OFF with a characteristic time, T, which is the inverse of the switching frequency. Also note that the characteristic drain current risetime and falltime is t_R and that T_D is the pulse width of the drain current. In order to achieve high-efficiency operation, the MOSFET should turn ON and OFF quickly.

The spectrum of the drain current includes a number of impulses at multiples of the switching frequency. The spectral envelope of the harmonics generated by this DC/DC converter is shown in Figure 7.6b. In Chapter 4, we showed that f_1 and f_2 are found by:

$$f_1 = \frac{1}{\pi T_d}$$

$$f_2 = \frac{1}{\pi t_r}$$

---

[5] We shall continue in this chapter to call high-frequency harmonics "interharmonics" to differentiate them from multiples of the line frequency.

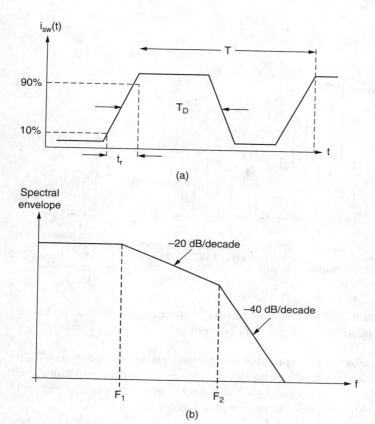

Figure 7.6 (a) The switching waveform of the drain current. The switching period is T, the pulse width is T_D, and the risetime of the current pulse is t_r. (b) The spectral envelope of the drain current.

Example 7.1: Spectrum of boost converter current. We'll find the spectrum of the drain current for a boost converter operating at a switching frequency $f_{sw} = 500$ kHz, a duty cycle $D = 0.5$, and with a switch risetime and falltime $t_r = 25$ nanoseconds. With a switching frequency of 500 kHz, the switching cycle $T = 2000$ nanoseconds, and a pulse width of $T_d = DT = 1000$ nanoseconds, we find the two corner frequencies as follows:

$$f_1 = \frac{1}{\pi T_d} = \frac{1}{(\pi)(1000 \times 10^{-9})} = 318 \text{ kHz}$$

$$f_2 = \frac{1}{\pi t_r} = \frac{1}{(\pi)(25 \times 10^{-9})} = 12.7 \text{ MHz}$$

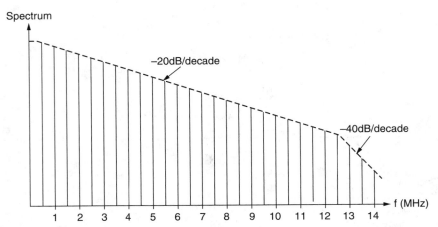

Figure 7.7   The spectrum of MOSFET drain current for Example 7.1. The spectral lines occur at multiples of 500 kHz. From 318 kHz to 12.7 MHz, the amplitude of the spectral lines falls off at –20 dB/decade. Above 12.7 MHz, the amplitude of the spectral lines falls off at –40 dB/decade.

Note that the fundamental switching frequency is 500 kHz, and that the overall spectrum of the switch current is as shown in Figure 7.7.

Example 7.2: Boost current spectrum with slower risetime and falltime.   We'll repeat the calculation of Example 7.2, now assuming the risetime and falltime is $t_r = 50$ nanoseconds. The first corner frequency has not changed, but the second corner frequency is now

$$f_2 = \frac{1}{\pi t_r} = \frac{1}{(\pi)(50 \times 10^{-9})} = 6.37\,\text{MHz}$$

This means the spectrum will be the same as Example 7.2 in the frequency range DC to 6.37 MHz. Above 6.37 MHz, the spectral components will fall off at a rate of –40 dB/decade. Slower switching of the MOSFET causes the higher frequency harmonics to roll-off faster, but incurs a penalty in power supply efficiency.

Testing for conducted EMI

Conducted EMI is the terminology used for the harmonic pollution that high-frequency circuits put onto the utility line. During a power supply design process, it is typical to test for conducted EMI in order to ensure compliance with EMI standards such as FCC Subpart J, or the CISPR standard EN55022. These standards are discussed in Chapter 2. A piece of equipment used during EMI testing is a line impedance stabilizing network (LISN), as shown in Figure 7.8.

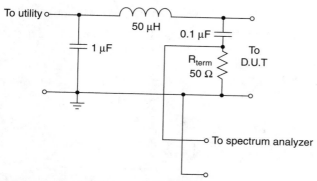

Figure 7.8 The LISN (Line Impedance Stabilization Network) connected between the utility and the device under test.

The LISN provides a controlled impedance (in this case 50 ohms) through which we measure the harmonic current injected onto the utility line by the device-under-test (D.U.T.). The 1 µF capacitor and the 50 µH inductor buffer the 50-Ω load from any noise on the utility line and force high frequency currents to flow through the 50-Ω resistor. We then measure the voltage across the resistor using a spectrum analyzer.[6] This measured voltage is proportional to the injected high frequency current that the D.U.T. would put onto the power line. FCC and CISPR standards set limits for the harmonic currents.

Corrective measures for improving conducted EMI

Most switching power supplies have an EMI filter on the front end, as shown in Figure 7.9a. This network shunts high-frequency currents from the switching power supply to ground (through C_2 and C_3), while allowing line frequency currents to pass through. Inductors L_1 and L_2 present high-impedance to high-frequency currents, further forcing conducted EMI to shunt to ground. In Figure 7.9b, we see a practical application of an EMI filter in the front end of a rectifier used in a wide-input-range power supply.

Summary

Switching power supplies are ubiquitous components in consumer and industrial equipment. In this chapter, we have seen that switching power supplies generate high-frequency voltages and currents, and that

---

[6] In this case, we'd connect the LISN to the spectrum analyzer using a 50-Ω BNC-type cable, with a 50-Ω termination at the spectrum analyzer.

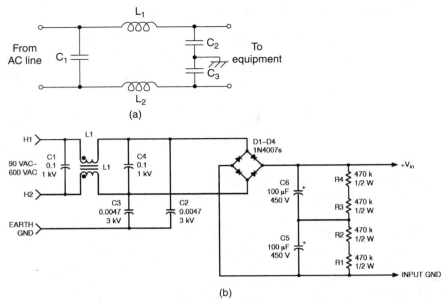

Figure 7.9   EMI filters for a single-phase application. (a) The basic circuit. (b) The input EMI filter and rectifier from the offline flyback supply [7.2].
[© 2002, On Semiconductor, reprinted with permission]

these high-frequency harmonics can adversely affect power quality by causing electromagnetic interference (EMI). The effects of EMI can be mitigated by the addition of proper high-frequency EMI filters.

References

[7.1] IEEE, "IEEE Recommended Practices and Requirements for Harmonic Control in Electrical Power Systems," IEEE Std. 519-1992, revision of IEEE Std. 519-1981.
[7.2] On Semiconductor, "Very Wide Input Voltage Range, Off-Line Flyback Switching Power Supply," Application Note AN1327/D, available from www.onsemi.com.
[7.3] ____, "MC33232 Power Factor Controller" datasheet, available from the Web: http://www.onsemi.com/pub/Collateral/MC33232.PDF.

Methods for Correction of Power-Quality Problems

In this chapter we discuss methods of correction of power-quality problems. Methods discussed include filters, transformers, uninterruptible power supplies (UPSs) and compensators.

Introduction

The first manifestation of a power-quality problem is a disturbance in the voltage waveform of the power source from a sine wave, or in the amplitude from an established reference level, or a complete interruption. The disturbance can be caused by harmonics in the current or by events in the supply system. The disturbance can last for a fraction of a cycle (milliseconds) to longer durations (seconds to hours) in the voltage supplied by the source. The disturbance usually becomes evident through mis-operation of the equipment supplied by the source, or because of a complete shutdown. Certain power-quality problems, such as current or voltage waveform distortion, will also appear as a result of a survey of a facility using appropriate instruments, as described in Chapter 14.

The objective of a "method for correction" is to make the power source meet a standard. For example, the standard can be for limits on rms voltage deviation in the form of an amplitude versus time chart, such as the classical CBEMA curve (as described in Chapter 2). The standard can be for voltage harmonic amplitudes in the form given in IEEE Std. 519. The standard also can be expressed as the duration of outages in the utility measure of SAIDI,[1] or the number of nines in the reliability per unit

---

[1] The SAIDI index is a standard measure of utility outage, and stands for "System Average Interruption Duration Index."

on-time, as described in Chapter 8.4. Power-quality standards are discussed in Chapter 2. Each problem may require a different correction method that will be discussed in this chapter, and in the following chapters.

Power-quality problems, particularly line-voltage disturbances, can originate at four levels of the system that delivers electric power, namely [8.1]:

- **Bulk power:** Power plants and the entire area transmission system
- **Area power:** Transmission lines, major substations
- **Distribution network:** Includes distribution substations, primary, and secondary power lines, and distribution transformers
- **Utilization equipment:** Includes service equipment and building wiring

In addition, the problems can be caused by the equipment supplied with electric power—for example, power-electronic converters. Redundancy at all levels of the electric-power system reduces the incidence and duration of line-voltage disturbances.

Correction Methods

Correction methods include the following:

- Design of load equipment
- Design of the electric-power supply system
- Installation of power-harmonic filters
- Use of dynamic voltage compensators
- Installation of uninterruptible power supplies (UPSs)
- Reliance on standby power—for example, engine-generator (E/G) sets

Load equipment, such as switch-mode power supplies, can be designed to reduce harmonics in the load current, and also to reduce sensitivity to voltage disturbances. The supply system can be designed to reduce source impedance, to separate loads, and to avoid harmonic resonances. Power-harmonic filters are installed to correct continuous voltage distortion produced by nonlinear loads in power systems, such as adjustable speed drives. Dynamic voltage compensators function to correct short-time voltage waveform sags. The available time duration of the correction depends on whether supplementary energy storage means, such as batteries, are incorporated in the compensator. A UPS provides an independent power source to the load from an electronic inverter. When utility power is interrupted, batteries serve as the energy source.

Engine-generator (E/G) sets increase the operating time of UPS beyond the available operating time of the batteries. The E/G sets serve for long-time utility outages, as well as the prime power supply when maintenance is required on the electrical system.

Correction methods for voltage disturbances are classified by whether or not they require a stored energy source, such as a battery, flywheel, fuel cell, or other means. As listed earlier, filters require no stored energy. Voltage compensators may require stored energy to handle deep and/or long-duration voltage sags. UPSs always require stored energy. E/G sets require fuel as the energy source.

Voltage disturbances versus correction methods

Before disturbances in power quality at a site can be corrected, the disturbance must be anticipated or identified. The objective of the correction must be established, and the correction method selected. The amplitude, waveform, and duration of voltage disturbances can be determined by measurement at the site, or by reports of typical disturbances made at the site or at other similar sites. The impact on the operation of equipment at the site is another measure of disturbances. An example of the distribution of voltage sags in low-voltage networks is shown in Figure 8.1 [8.3].

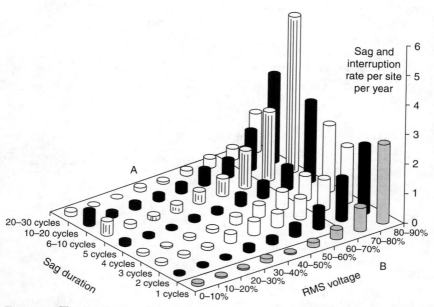

Figure 8.1 The distribution of sag and interruption rates in low-voltage networks in the U.S. [8.3].

The distribution of voltage sags shown in Figure 8.1 can be compared with the ITIC Equipment Sensitivity curve in Figure 8.2. Equipment built to operate within the voltage tolerance envelope is supposed to operate without interruption for voltage sags within the given amplitude and time duration envelope. For sags in Region A, namely 1 cycle to 30 cycles and 70 percent remaining voltage, the equipment will be out of the envelope and may be interrupted. However, the interruption rate shown in Figure 8.1 for Region A is one per site per year. For sags in Region B, namely zero voltage for up to 1 cycle, and 70 percent remaining

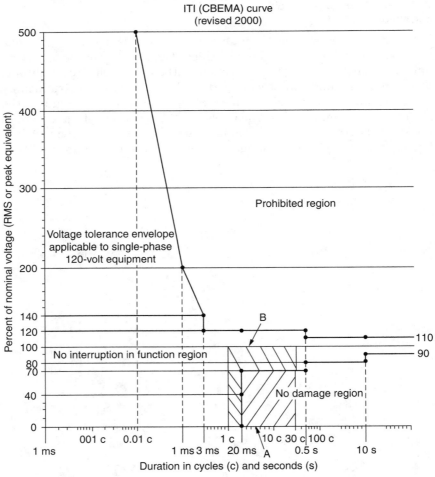

Figure 8.2   An ITIC equipment sensitivity curve. Regions A and B correspond to the distribution in Figure 8.1 [8.3].

voltage for 1 cycle to 30 cycles, the equipment should not be interrupted, even though the rate of sags can be up to six per site per year.

The objective of the correction methods has been discussed in Chapter 2 in the form of voltage versus time curves defined as CBEMA and ITIC. The theory being that if the source voltage is corrected to the acceptance regions of these curves, then equipment designed to operate in these regions will operate without interruption with the corrected voltage. An introduction to correction methods is given in Table 8-1. Correction methods are listed for typical voltage disturbances and time durations.

Manufacturers offer variations of the end-use equipment, which can withstand ranges of voltage sag and time duration. Furthermore, protected loads, like computers, can be divided and supplied by multiple correction sources—for example, UPSs [8.3].

Reliability

The reason for correcting power-quality problems is to insure the reliability of the equipment supplied by electric power from the system in which the problems occur. Although specific equipment, such as a battery-inverter UPS, is given reliability numbers, like mean-time to failure (MTBF), data centers supplied by multiple UPSs are classified in terms of availability by using tier designations, as follows [8.5]:

- **Tier 1:** Tier I is composed of a single path for power and cooling distribution, without redundant components, providing 99.671 percent availability. (Unavailability, 28.8 hours per year.)

- **Tier II:** Tier II is composed of a single path for power and cooling distribution, with redundant components, providing 99.741 percent availability. (Unavailability, 22.7 hours per year.)

- **Tier III:** Tier III is composed of multiple active power and cooling distribution paths, but only one path active, has redundant components, and is concurrently maintainable, providing 99.982 percent availability. (Unavailability, 1.58 hours per year.)

- **Tier IV:** Tier IV is composed of multiple active power and cooling distribution paths, has redundant components, and is fault tolerant, providing 99.995 percent availability. (Unavailability, 0.44 hours per year.)

The ultimate availability is termed, "five nines," or 99.999 percent availability. The unavailability is 5.26 min per year.

Additional descriptions of equipment for data centers will be described in Chapter 9.

TABLE 8.1 Voltage Disturbances versus Correction Methods

Voltage disturbance	Duration	Correction method	Maintenance required	Energy storage	Limitations	Cost $/kVA
1. Waveform distortion	Continuous	Power-harmonic filters	No	No	None/tuning	
2. Surge to 200%	0.17 ms	Surge suppressor	No	No	Aging	50–200
Surge to 200%	1 ms	Surge suppressor	No	No	Aging	50–200
3. Sag to 50% at 50% load	Continuous	Const.-volt transformer	No	No	Size	40–1500
Sag to 70% at 100% of load	Continuous					
4. Sag to 50%	0.2 s	Volt compensator	No	No	None	—
Sag to 0%	2 s			Yes	Batteries	—
5. Sag to 0%	1–60 min	UPS	Yes	Yes	Batteries	1000–3000
6. Outage	Over 60 min	UPS plus E/G set	Yes	Yes	Batteries/fuel	1250–3500

Design of load equipment

Two factors in the design of load equipment can (1) reduce the possibility of the equipment itself causing a power-quality problem, such as producing harmonic currents, and (2) reduce the sensitivity of the equipment to problems such as voltage sags and outages.

Six-pulse rectifiers, which serve as the front end of ASDs and UPSs, distort the line currents, as described in Chapter 5. Two remedies can be employed: (1) twelve-pulse rectifiers, and (2) pulse-width modulation (PWM) of the line current. A twelve-pulse rectifier circuit is shown in Figure 8.3a [8.6]. The two six-pulse bridges are supplied from delta and wye secondary windings of the supply transformer to obtain the 30-degree phase shift between the source voltages to the rectifier bridges. The resultant line currents are shown in Figure 8.3b. The 5th and 7th harmonics are eliminated; the lowest order harmonic is now the 11th. The effect of PWM in the line current by switching the devices in a six-pulse rectifier is shown in the waveforms of Figure 8.4 [8.7].

Equipment subject to source voltage sags will respond in one of the following ways:

- Restart with no damage—for example, home appliances.

- Restart with some damage—for instance, computer with damage to functions that prevent a restart.

- Require manual intervention to restart—such as motors in equipment where automatic restarting may be a hazard.

- Will not restart—for example, the equipment is damaged due to voltage sag.

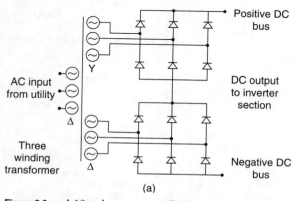

(a)

Figure 8.3a A 12-pulse converter. The bridges are connected in series [8.6].
[© 1986, IEEE, reprinted with permission]

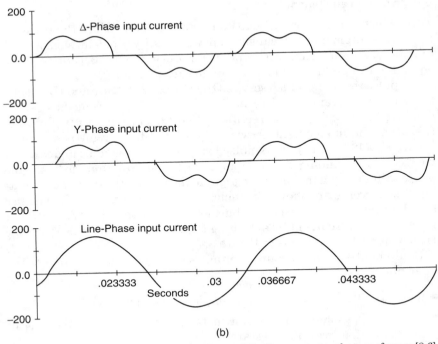

Figure 8.3b   The input current to each bridge, and the line current to the transformer [8.6]. [© 1986, IEEE, reprinted with permission]

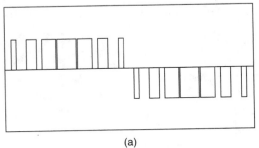

(a)

Figure 8.4a   The pulse-width modulation (PWM) of a rectifier [8.7].

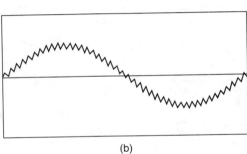

(b)

Figure 8.4b   The resultant line current from PWM [8.7]. [© 1997, IEEE, reprinted with permission]

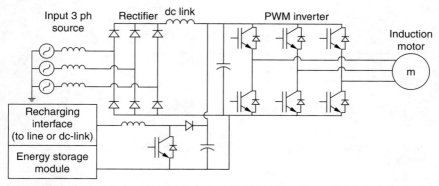

Figure 8.5   The energy storage module to provide ride-through capability for ASD during line-voltage sags [8.8].
[© 1999, IEEE, reprinted with permission]

The sensitivity to voltage sags can be mitigated by either supplying the critical equipment from a battery-inverter UPS or improving the specific parts of the equipment responsible for the sensitivity. Figure 8.5 shows an example of a supplementary power source to the DC link of the rectifier-inverter of an ASD to assist it in withstanding voltage sags [8.8].

**The design of electric-power
supply systems**

The correction methods for the following two types of power-quality problems that depend upon the design of the system include the following:

- Harmonic currents
- Voltage sags and surges

The design of an electric-power system subject to harmonic currents must prevent the effects of the currents, namely:

- Raising the rms current levels in conductors, transformers, and capacitors
- Causing resonances, which result in excessive capacitor and system voltages

The analysis for the effect of harmonics is treated in the reference IEEE Std. 519-1992 [8.15]. An illustration of how a system is modeled is shown in Figure 8.6. The harmonic currents are assumed to originate

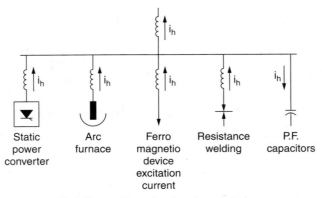

Figure 8.6 Modeling nonlinear loads as harmonic current sources per IEEE Std 519-1992 [8.15].
[© 1992, IEEE, reprinted with permission]

in the nonlinear loads—for example, the static power converter—and return in the capacitors and supply system.

To calculate the currents and voltages, the system is reduced to the form shown in Figure 8.7 for each harmonic current i_h. For example, for the fifth harmonic value of i_h, the reactances of X_L and X_C are calculated for 300 Hz. Obviously, resonance will occur when the inductor reactance X_L equals the capacitive reactance X_C at or near the harmonic frequency.

The design of an electric-power system to reduce the effect of voltage sags and surges that originate within the facility includes the following steps:

- Insure the low impedance connection of loads to the power source.
- Isolate loads from disturbances.
- Utilize ample transformer and conductor sizes.
- Switch power-factor correction capacitors in small steps.
- Utilize soft motor starters.

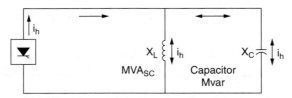

Figure 8.7 An equivalent circuit for calculating the effect of harmonic current ih [8.15].
[© 1992, IEEE, reprinted with permission]

These steps do not include the steps the utility can take to minimize the sags and surges that occur in the source voltage supplied by the utility to the facility.

Power harmonic filters

Power-harmonic filters are inserted into a power system to absorb specific harmonic currents generated by nonlinear loads such as the converters of ASDs. The filters can be passive, tuned to a fixed frequency that is usually slightly below the harmonic frequency. The filters can be built as active filters—that is, to be tunable to account for changes in the system impedances and loads. The subject of power-harmonic filters is addressed in Chapter 6.

Utilization-dynamic voltage compensators

The dynamic voltage compensator corrects voltage sags by inserting a voltage component between the power source and the load to maintain the required load voltage. The power for the correction is usually taken from the source, but supplementary energy storage is sometimes used. The inserted voltage component is shaped in amplitude and waveform by a controller. Compensation is usually limited to 12 cycles for dips to zero source voltage, and to 2 seconds (s) for dips to 50 percent source voltage. One circuit concept is shown in Figure 8.8 [8.9]. Because it acts for a short time and has no energy storage, the compensator is smaller and lower in cost than a battery-inverter UPS. However, it cannot compensate for long-term outages—for example, a duration of minutes. The subject is treated in Chapter 10.

Uninterruptible power supplies

The most commonly used equipment to protect critical loads from power-quality problems is the battery-inverter UPS. The concept is shown in Figure 8.9 [8.10]. The basic parts of the module are the battery, the inverter, and the input rectifier, which also serves as the battery charger.

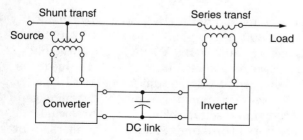

Figure 8.8 A dynamic voltage compensator.

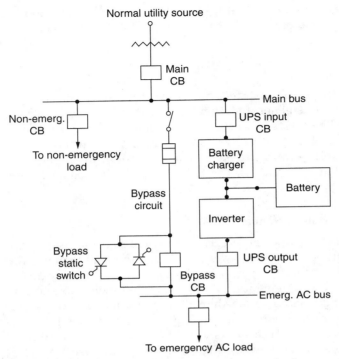

Figure 8.9   An electrical system for non-emergency load, emergency (critical) load by battery-supported UPS, and bypass circuit [8.10].

In addition, a high speed bypass switch is incorporated to provide power to the load if the inverter fails. The output of the UPS module is independent of power-quality problems in the supply system, and only limited by the ampere-hour capacity of the battery or an E/G set.

Battery-powered UPS modules are available in power levels of 100 W to 500 kW. The modules can be operated in systems for higher ratings—for example, up to 10,000 kW. Details are offered in Chapter 9.

Transformers

Transformers provide service to, and within, a facility from the utility source, typically 13.8 kV to 480 V, three-phase; and for utilization voltages within a facility, typically 480/277/120 V, single and three-phase. These transformers can also correct power-quality problems due to harmonic currents. In addition, constant voltage transformers utilizing ferroresonance are used to correct for short-term and long-term sags in source voltage down to a remaining voltage of 70 percent for local loads. Utility substation transformers utilize under-load tap changers to correct for slow deviations in voltage [8.11].

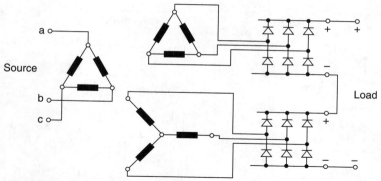

Figure 8.10 A 12-pulse rectifier supplied by a three-winding transformer with delta and wye secondary windings.

A common application of transformers is the input to a 12-pulse rectifier, as shown in Figure 8.10. The transformer has a wye and delta secondary to supply each of the rectifier bridges. The 30-degree phase shift between the two secondary voltages serves to cancel the fifth and seventh harmonics of the primary current. Another use of transformers is to cancel harmonics, as shown in Figure 8.11. The two secondary windings are zig-zag to provide 30 degrees of phase shift between secondary line-to neutral voltages for the two sets of equal-power loads. The same effect can be achieved with wye-delta secondary windings with half the loads operating at 120 V and half at 277 V, single-phase. A third example

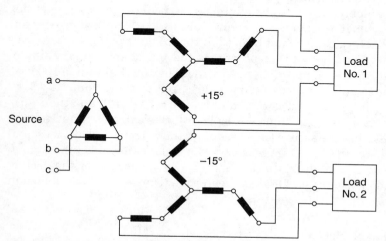

Figure 8.11 Harmonic cancellations for two equal loads using a transformer with ±15-degree zig-zag secondary windings.

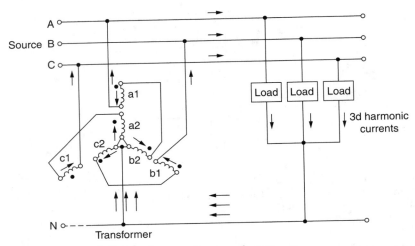

Figure 8.12   A third harmonic grounding transformer.

is the grounding transformer shown in Figure 8.12 [8.12], which is located close to a collection of single-phase loads that produce 3d-harmonic currents. The third-harmonic currents are prevented from traveling back to the source in the neutral conductor.

The constant voltage transformer (CVT) utilizes a leakage-reactance transformer with a saturable magnetic core and a capacitor to obtain relatively constant output voltage in the face of input voltage sags down to a remaining voltage of 70 percent at full load and to 30 percent at a 25 percent load. The characteristic curves for time in cycles and percent load are shown in Figure 8.13 [8.13]. The response occurs within a half cycle and is not limited in time. The transformers are relatively large and heavy, typically twice the size of an ordinary single-phase transformer of the same kVA rating. They are built up to 100 kVA single-phase and can be banked for three-phase operation.

The circuit for the CVT is shown in Figure 8.14 [8.13], and the equivalent circuit appears in Figure 8.15 [8.14]. Basically, as input voltage V_o declines, the core saturation is reduced, the current I_m decreases, and the net current ($I_c - I_m$) increases through the series reactance, causing the output voltage V_m to rise. The rise compensates for the decline in V_o [8.14]. Structures of leakage reactance transformers are shown in Figure 8.16 [8.14].

Standby power systems

The battery-powered UPS can only deliver power to its critical load for the ampere-hour discharge time of the batteries. The battery capacity is

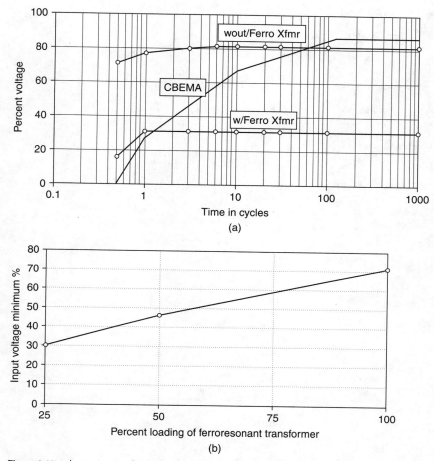

Figure 8.13 A constant-voltage ferroresonant transformer [8.13]. (a) A protecting single-loop process controller with and without ferroresonant transformer against voltage sag. (b) Minimum regulated voltage versus percent loading.

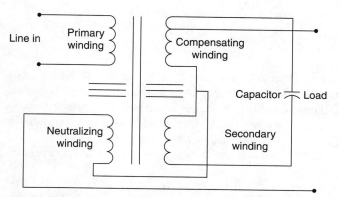

Figure 8.14 Schematic diagram of a constant-voltage ferroresonant transformer [8.13].

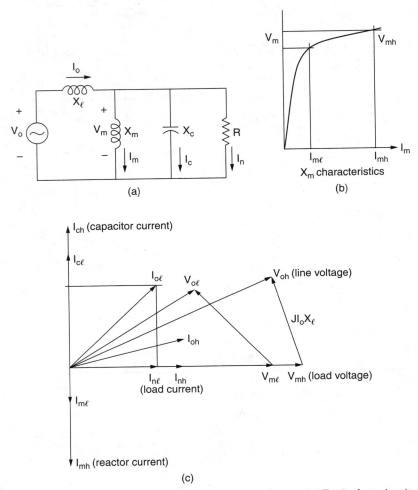

Figure 8.15 The operation of a ferroresonant transformer. (a) Equivalent circuit. (b) Characteristics of shunt reactor. (c) Phasor diagram for high and low line voltage [8.14].

a cost and space issue. Typical times are in the range of 3 to 10 minutes for the UPS delivering rated power. For longer operating times without utility power, a UPS needs a standby engine-generator (E/G) set to supply power and recharge the batteries. A simplified diagram is shown in Figure 8.17 [8.10]. The controls for the transfer switch detect the loss of utility power, start the E/G set, and initiate the transfer of the emergency load (UPS) to the E/G set.

Large systems requiring continuous power supply from UPSs are termed data centers. In addition to electric power for critical loads,

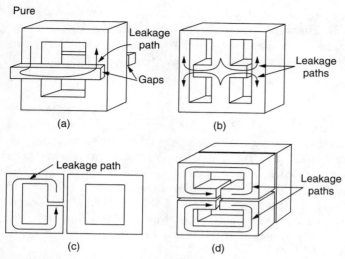

Figure 8.16 Leakage-reactance transformer core construction: (a) bar shunt; (b) X core; (c) single leakage core; and (d) double leakage core [8.14].

such as computers, these centers require power for supporting functions like air conditioning, lighting, heat exchanging, living facilities, and maintenance. The support is usually provided by E/G sets or alternate utility feeders. The centers must also be designed so equipment

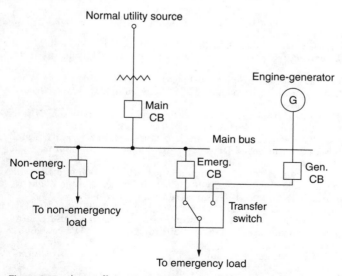

Figure 8.17 A standby system with engine generator and transfer switch [8.10].

can be maintained or replaced without interrupting power supply to critical loads [8.4]. The description of reliability includes the Tier concept for definitions of availability time, while Chapter 13 provides details of standby power systems.

Summary

Deviations in source voltage and current for critical load equipment must be corrected to insure reliable operation of the equipment. Industry standards, such as CBEMA, define corrected values. Methods of correction include filters, compensators, special transformers, and battery-inverter UPSs. The UPS modules are utilized singly and in groups as a function of the requirements of the load. Engine-generator sets serve to extend the operating time of batteries in UPS. Reliability of systems supported by UPS is measured in percent availability, from Tier I to Tier IV.

References

[8.1] "Reliability Models for Electric Power Systems," whitepaper #23, American Power Conversion (APC), 2003.

[8.2] A. T. de Almedia, F.J.T.E. Ferreira, and D. Both, "Technical and Economical Considerations in the Application of Variable-Speed Drives with Electric Motor Systems," IEEE Transactions on Industry Applications, vol. 41, no. 1, Jan./Feb. 2005, pp. 188–199.

[8.3] M. Andressen, "Real Time Disturbance Analysis and Notification," Power Quality Conference '04, Chicago, IL.

[8.4] R. J. Yester, "New Approach to High Availability Computer Power System Design," EC&M, January 2006, pp. 18–24.

[8.5] W. P. Turner IV, J. H. Seader, and K. G. Brill, "Industry Standard Tier Classifications Define Site Infrastructure Performance," whitepaper, The Uptime Institute, 2001–2005.

[8.6] J. W. Gray and F. J. Haydock, "Industrial Power Quality Considerations When Installing Adjustable Speed Drive Systems," IEEE Trans. on Ind. Appl., vol. 32, no. 3, May/June 1986, pp. 646–652.

[8.7] L. Manz, "Applying Adjustable-Speed Drives to Three-Phase Induction NEMA Frame Motors," IEEE Trans. on Ind. Appl., vol. 33, no. 2, March/April 1997.

[8.8] A. von Jouanne, P. N. Enjeti, and B. Banerjee, "Assessment of Ride-Through Alternatives for Adjustable Speed Drives," IEEE Trans. on Ind. Appl., vol. 35, no. 4, July/August 1999, pp. 908–916.

[8.9] T. Jimichi, H. Fujita, and H. Akagi, "Design and Experimentation of a Dynamic Voltage Restorer Capable of Significantly Reducing an Energy-Storage Element," Conference Record, 2005 Fortieth IAS Annual Meeting, pp. 896–903.

[8.10] A. Kusko, Emergency Standby Power Systems, McGraw-Hill, New York, 1989.

[8.11] J. G. Boudrias, "Harmonic Mitigation, Power Factor Connection, and Energy. Saving with Proper Transformers and Phase Shifting Techniques," Power Quality Conference, '04, Chicago, IL.

[8.12] L. F. Blume, G. Camilli, A. Boyajian, and V. M. Montsinger, Transformer Engineering, John Wiley, 1938.

[8.13] R. C. Dugan, M. F. McGranaghan, S. Santosa, and H. W. Beaty, Electrical Power Systems Quality, 2nd edition, McGraw-Hill, 2002.

[8.14] A. Kusko and T. Wrobleski, *Computer-Aided Design of Magnetic Circuits*, The M.I.T. Press, 1969.

[8.15] IEEE, "IEEE Recommended Practices and Requirements for Harmonic Control in Electrical Power Systems," IEEE Std 519-1992.

[8.16] A. Kusko and S. M. Peeran, "Application of 12-Pulse Converters to Reduce Electrical Interference and Audible Noise from DC-Motor Drives," 1989 IEEE-IAS Annual Meeting, San Diego, October, 1989.

Uninterruptible Power Supplies

The term Uninterruptible Power Supplies (UPS) has been applied both to an uninterruptible power system and to the specific battery-inverter equipment incorporated into the system. In this chapter, we will use the term UPS for the specific equipment.

The UPS was developed in parallel with digital computers and other IT equipment to provide a reliable uninterruptible electric power source to create a standard the electric-utility industry could not provide. The computer industry would not have developed without some type of UPS, which either used engine-generator sets or static inverters employing power electronic devices. If the concept of an independent UPS had not been developed, then all electronic equipment sensitive to disturbances in source voltage would have had to incorporate energy storage means—for example, batteries to counteract such disturbances in the source voltage.

Introduction

The concept of a double-conversion UPS is shown in Figure 9.1a [9.1]. The essential parts are

- The **inverter**, for converting power from DC to AC
- The **battery**, to supply power to the inverter when AC power is interrupted
- The **converter-battery charger**, to supply DC power to the inverter, as well as to charge the battery
- The **by-pass circuit**, to supply AC power directly to the load when the inverter fails. A second by-pass circuit external to the UPS—the

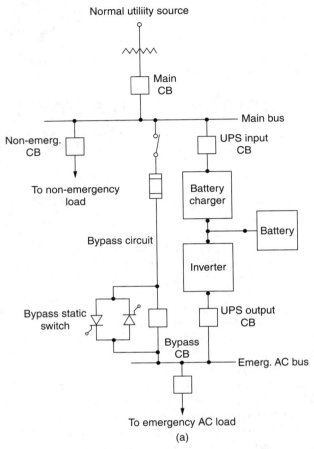

Figure 9.1a Double conversion UPS: inverter; battery; converter-battery charger; by-pass circuit [9.1].

maintenance by-pass—is usually provided when the UPS is taken out of service.

Both the battery charger (rectifier) and the inverter utilize power semiconductor devices, which are switched to convert power from AC to DC, or DC to AC. The theory of operation is described in Chapter 5. The devices are usually pulse-width modulated to shape the line-current waveform of the rectifier, as well as the output voltage of the inverter to sine waveforms. Additional elimination of input and output harmonics is done with filters.

The concept of the "Delta UPS" is shown in Figure 9.1b. The battery-inverter circuit only supplies the "delta" or difference between the line

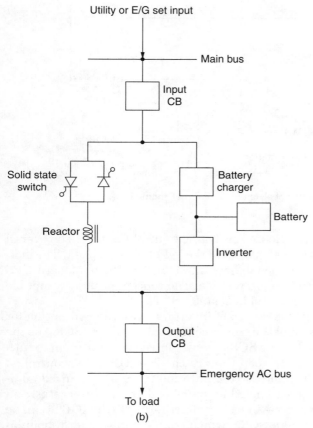

Figure 9.1b Delta UPS: solid-state switch; reactor; battery charger; inverter; battery.

voltage and the required load voltage. The Delta UPS is more efficient than the double-conversion UPS because the load power is supplied directly from the line nearly all of the time. The inverter and battery still have to be sized for the longest utility outage time.

History

The silicon-controlled rectifier (SCR) was introduced in 1957. The development of inverters using SCRs for UPS was expedited by the book, *Principles of Inverter Circuits*, by Bedford and Hoft, published in 1964 [9.2]. An early publication by Fink, Johnston, and Krings in 1963 described three-phase static inverters for essential loads up to 800 kVA [9.3]. Kusko and Gilmore described in 1967 a four-module redundant UPS for

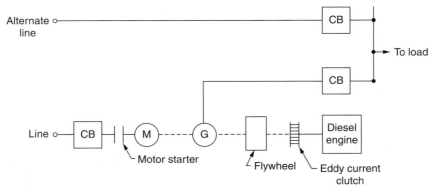

Figure 9.2a MG UPS with flywheel, clutch, and diesel engine [9.1].

500 kVA for the FAA Air Route Air Traffic Control Centers. The overall MTBF requirement was 100,000 h and the MTTR was 1h [9.4].[1] The individual modules were assumed to have an MTBF of 1100 h. Modern UPS modules are considerably more reliable. The concept of uninterruptible system availability is described in Chapter 8.

The major change in the design of inverters since their inception for UPSs is the replacement of the silicon controlled rectifiers (SCRs) as the switch devices by IGBTs. The SCRs require forced commutation to turn off the current—that is, to open the switch. The IGBTs are controlled by the gate voltage. The forced commutation for SCRs required additional circuits and operations that increased the failure rates and reduced the reliability of the inverters. Furthermore, the IGBTs can be switched much faster than SCRs, which enabled pulse-width modulation (PWM)—in other words, the voltage and current waveform shaping to be used.

An early concept of UPS that preceded the SCR is shown in Figure 9.2a. It was a rotary machine set consisting of a diesel engine, clutch, electric motor, flywheel, and generator [9.1]. When utility power was available, the motor drove the generator. When utility power failed, the engine would be started while the flywheel and the generator slowed. At a certain speed, the clutch was engaged so the engine could drive the generator to restore and maintain its synchronous speed. A motor-generator which is a UPS, but employs batteries to provide energy when utility power fails, is shown in Figure 9.2b.

---

[1] MTBF = Mean time between failures. MTTR = Mean time to repair.

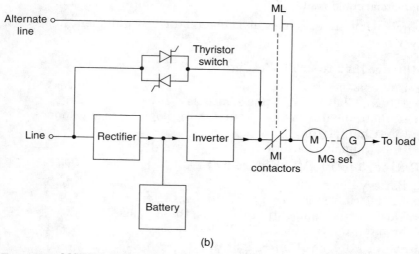

Figure 9.2b   MG UPS with load-commutated inverter and battery [9.1].

Types of UPS Equipment

Commercially available UPS equipment can be categorized into the following types, depending upon the manufacturer and the application.

- **Output Ratings**
  - Power: 1 kVA single-phase to 1000 kVA three-phase per module
  - Voltage: 120 V single-phase to 208 V/120 V or 480 V three-phase
- **Design**
  - Double: Conversion or Delta
  - Inverter: Power transistors, IGBTs. Pulse-width modulated with output filter
  - Battery Charger: Input rectifier or separate charger
  - By-pass Circuit: Thyristors or IGBTs
  - Maintenance By-pass: Circuit breaker
- **Energy Storage**
  - Batteries: Flooded, VLRA (value regulated lead acid)
  - Flywheels
  - Fuel cells
- **Typical Run Time**
  - Batteries: 5 to 20 minutes at full load
  - Flywheels: 15 seconds, enough time to start E/G set
  - Fuel cells: 8 hours

Commercial equipment

The following are examples of commercial UPS equipment in order of power ratings:

- **Rating, 1000 to 2200 VA (Figure 9.3)**
  - Delta battery-inverter
  - 120- and 240-V single-phase models
  - AVR, automatic voltage regulator, topology, voltage compensation (See Chapter 10)
  - Hot swappable batteries
- **Rating, 3 to 10 kVA (Figure 9.4) [9.5]**
  - Battery-inverter
  - Double-conversion (online rectifier-inverter)
  - Various input and output voltages, three-phase
  - Monitor and control via the Web browser
- **Rating, 120 kVA (Figure 9.5) [9.6]**
  - Battery-inverter
  - 480 V, three-phase
  - Run-time: full load, 8 minutes; half load, 22 minutes
  - Interphase port, DB-25, RS-232
- **Rating, 1000 kVA, (Figure 9.6) [9.6]**
  - Battery inverter
  - 480 V, three-phase
  - Configurable for N+1 internal redundancy

The slim 2U shape and advanced networking and remote control options make this UPS the perfect choice for networking or process control racks. The S3K2U series UPS's protect against most severe power disturbances including over/under voltages through state of art sinewave, line-interactive technology. Utility power is continually protected by the S3K2U series while internal battery power is maintained for deep sag conditions.

The built-in protection for under and over voltage conditions of the S3K2U units includes low-energy lightning surge protection on the input power source. They are provided with an input circuit protector and a pair of surge protected data line connectors (RJ-45).

The S3K2U series include an automatic bi-weekly test function to ensure the capability of the battery to supply power in deep sag situations. Should the battery fail this test, the UPS will display a warning to indicate that the battery needs to be replaced.

Figure 9.3 Delta line-interactive UPS, 1000VA to 2200 VA. [Courtesy Sola/Hevi-Duty]

Industrial UPS

The double-conversion online Allen-Bradley 1609-P UPS is available in the 3 kVA to 10 kVA power range. Each model is available in various input and output voltage combination and offers assorted output voltage and receptacle or hard wired configurations. Its network management card allows users to monitor and control capabilities via a standard Web browser or RSView.
Rockwell Automation

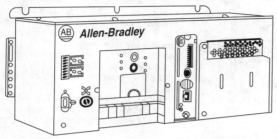

Figure 9.4 Double-conversion UPS, 3 to 10 kVA [9.5].

- ■ Modular design for fast service and reduced maintenance requirements
- ■ **Rating, for 225 kVA UPS, (Figure 9.7) [9.7]**
  - ■ Flywheel power system support for UPS
  - ■ Run time, 13 seconds at UPS full load, 28 seconds at UPS 50 percent load
  - ■ Recharge time, 20 seconds
  - ■ N+1 configurable

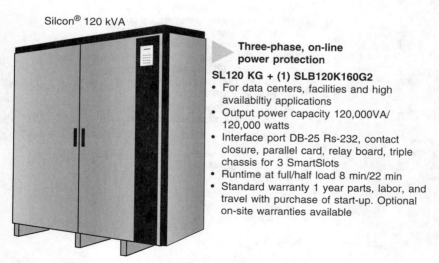

Silcon® 120 kVA

Three-phase, on-line power protection

SL120 KG + (1) SLB120K160G2
- • For data centers, facilities and high availabiltiy applications
- • Output power capacity 120,000VA/ 120,000 watts
- • Interface port DB-25 Rs-232, contact closure, parallel card, relay board, triple chassis for 3 SmartSlots
- • Runtime at full/half load 8 min/22 min
- • Standard warranty 1 year parts, labor, and travel with purchase of start-up. Optional on-site warranties available

Figure 9.5 Double-conversion UPS, 8 min run time, 120 kVA [9.6].
[Courtesy APC]

Symmetra® MW 1000 kVA

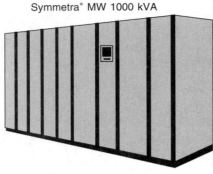

The world's largest modular UPS

Sy1000K1000G + SYMBP1000C1G12 + (3)SYB400K1000GXR-2C

- 1000 kVA/kW 480 V UPS
- Scalable power capacity reduces UPS oversizing costs
- Configurable for N + 1 internal redundancy provides high availabilty
- Modular design for fast service and reduced maintenance requirements
- LCD display provides schematic overview of critical data
- Fully rated power kVA equals kW. This reduces cost by eliminating the need for an oversized UPS for Power Factor Corrected (PFC) loads

Figure 9.6   Double-conversion UPS, 1000 kVA [9.6]. [Courtesy APC]

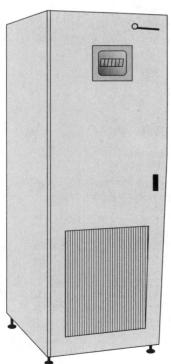

Flywheel power system

The VSS+ voltage support solution (VSS) flywheel power system provides ride-through protection for a safe system shutdown of most process operations or until a standby engine-generator can come online. It also handles short-duration power disturbances so UPS batteries can be saved for longer events. A single VSS+ unit provides up to 25 sec. in a 130 kVA UPS and 40 sec. in a 80 kVA UPS. For large systems, multiple VSS+ system can be paralleled together without any additional communication links. *Pentadyne*

Figure 9.7   Flywheel UPS, 130 kVA, 25 s [9.7]. [Courtesy Pentadyne Power Corporation]

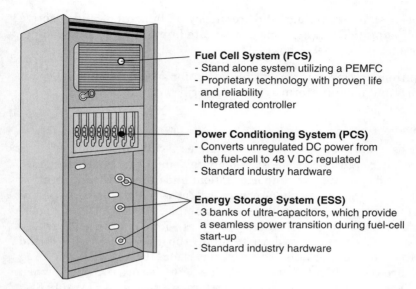

Fuel Cell System (FCS)
- Stand alone system utilizing a PEMFC
- Proprietary technology with proven life and reliability
- Integrated controller

Power Conditioning System (PCS)
- Converts unregulated DC power from the fuel-cell to 48 V DC regulated
- Standard industry hardware

Energy Storage System (ESS)
- 3 banks of ultra-capacitors, which provide a seamless power transition during fuel-cell start-up
- Standard industry hardware

Figure 9.8   Fuel-cell UPS, 48 V DC power, 4.5 kW [9.8].
[Courtesy UTC Power]

- **Rating, 4.5 kW, (Figure 9.8) [9.8]**
  - Fuel cell–powered UPS
  - Ultra capacitors for fuel cell startup
  - Output, 28 V DC

Energy storage

The requirement for stored energy in an uninterruptible (standby) power system is predicated on at least two parameters: (1) the time duration of power delivery (term), and (2) the power level (energy). The requirements further defined by the time duration can be categorized as follows [9.8]:

- **Short Term:** Standby systems without available transfer means to engine-generator sets or alternate utility feeders. These stand-alone systems range from 100 W to 1000 kW and include 5 to 30 minutes of stored energy capability, based on estimates of utility outage time.

- **Medium Term:** Standby systems with available transfer means to engine-generator sets, alternate utility feeders, or other sources. These systems range up to 10,000 kW and include up to 5 minutes of stored energy capability, based on the time to start engine-generator sets and make the transfer.

- **Long Term:** Standby systems that operate as part of a utility system, which provide, in addition to standby function, other functions such as peak shaving, voltage and frequency stabilization, and reactive power supply. These systems can be rated up to 20 MW and can deliver energy for up to 8 hours.

Requirements for uninterruptible power for specific loads can be met by short- and medium-term systems described earlier.

Batteries

Batteries consist of one or more cells electrically interconnected to achieve the required voltage, stored energy, and other characteristics. Two types of operation are important: float and cycling. Float operation describes batteries in telephone central offices where the batteries maintain a relatively constant voltage—for example, 48 V DC. Cycling operation describes batteries in standby systems—for example, UPS, where the battery charge is drawn down to supply the inverter and the AC load when the utility power fails. These batteries for UPS rated 100 kVA and higher, are typically rated 460 V DC. The batteries are recharged when utility power returns, or engine generators are started and run [9.8].

The specific energy and energy density of the batteries used for standby service are shown in Figure 9.9 [9.9]. The application of these batteries depends on additional factors besides those in the figure. The batteries employed for standby service are described in the following [9.8]:

- **Flooded, lead acid batteries:** These have been used for UPSs since the 1960s [9.4], and as the backup for communications power supplies before 1983 [9.10]. This type of battery requires periodic additions of water to comply with its specific gravity measurements. It discharges inflammable gas, and thus requires special facilities for safety. To facilitate venting, the gas space in flooded cells is open to outside air but

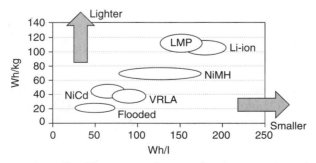

Figure 9.9 Specific energy and energy density comparison of batteries: Wh/kg and Wh/l [9.9].
[© 2004, IEEE, reprinted with permission]

separated from it through a vent that incorporates a flash arresting device. Note in Figure 9.9 that the flooded lead acid battery has the lowest specific energy and lowest energy density compared to other batteries.

- **Valve-regulated lead-acid (VRLA) batteries:** These have seen tremendous growth in standby usage in the last two decades [9.11]. Note their approximately two-to-one advantage over flooded batteries in Figure 9.9 in specific energy and energy density. In the VRLA cell, the vent for the gas space incorporates a pressure relief valve to minimize the gas loss and prevent direct contact of the headspace with the outside air.

Standard VRLA battery warranties range from 5 to 20 years depending upon their construction, manufacturer-based requirements concerning proper charging and maintenance, and whether the battery is kept in a 25°C (77°F) environment compared to a −40 to +65°C outdoor environment. When placed in an outdoor environment, the batteries must be heated to prevent freezing, or loss of capacity. At −6°C (20°F), battery capacity is reduced by 30 percent. At −16°C (4°F), battery capacity is reduced by 55 percent. [9.12]

Flywheels

Flywheels were the original means for energy storage in early designs of "no-break" engine-generator sets. (See Figure 9.2a.) They are returning to serve for short-time supply in standby systems as an alternative to batteries, and in other applications.

The energy stored in a flywheel is given by the classical equation:

$$W = (1/2)I\omega^2$$

where W = energy, joules or watt seconds ($m^2 \times kg/s^2$)
    I = moment of inertia ($N \times m \times s^2$)
    ω = rotational velocity (rad/s)

Note that the energy W stored in the flywheel is always known by the speed ω.

Sample ratings are given by Weissback [9.13] of low speed systems (less than 10,000 rpm) capable of delivering power over 1 MVA, with energy storage below 10 kWh. Reiner [9.14] describes a flywheel plant concept that can supply power peaks of 50 MW for about 13 s, equivalent to energy storage of 181 kWh.

For perspective, consider a UPS that requires 1000 kW at its DC bus for 10 s to insure time for start up and transfer to back-up engine generators. The calculated energy is 2.78 kWh. Assume that the flywheel speed slows to 70 percent and that the flywheel generator and converter efficiency is 0.90, the calculated flywheel stored energy must be 6.3 kWh at full speed.

Applications. Applications of flywheels include the following:

- No-break engine-generator set with flywheel and clutch [9.15]. The flywheel provides energy to the generator when the utility source fails, until the engine starts and reaches operating speed.

- AC UPS, which delivers AC power to the load when the utility source fails, as shown by Lawrence in Figure 9.10 [9.15].

- Battery substitute in UPS, as shown by Takashi in Figure 9.11 [9.16].

- Support medium voltage distribution network, as described by Richard against voltage sags and interruptions [9.17].

Profactors primarily comparing flywheels to batteries include the following:

- Maintenance-free, bearings might need service in three to five years [9.18]. Bearing-free flywheels utilizing magnetic levitation have been built.

- Long life—for example, three times that of batteries [9.16].

- Can provide typically 15 s of power for engine, or turbine-generator start [9.18].

- Short recharge time; depends on power available—for example, for one-tenth the discharge power, approximately 20 times the discharge time [9.19].

- Smaller footprint than batteries [9.19].

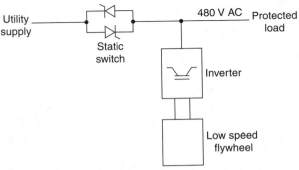

Figure 9.10 Block diagram of flywheel UPS with static switch, inverter [9.15].
[© 2003, IEEE, reprinted with permission]

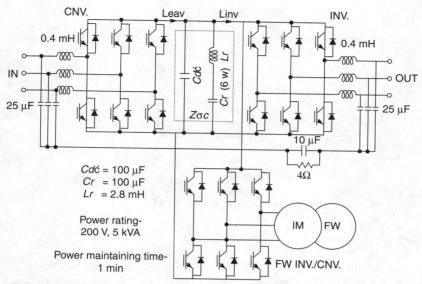

Figure 9.11 Electrical diagram of double-conversion UPS with flywheel energy storage, rating 5 kVA, 1 min [9.16].

- Minimum end-of-life disposal problem [9.16 and 9.19].
- Ambient temperature (0 to 40°C) compared to batteries (20 to 25°C) [9.19].
- Measure available energy accurately by calculating speed and energy.
- Can provide AC generator or DC converter output.

Con factors include the following:

- Installed cost 1 to 1.4 times that of batteries [9.19]
- Storage expansion not easy, requires adding units of comparable size

Other pro and con factors include availability and operator's experience with flywheels.

Fuel cells

Fuel cells, using hydrogen as a fuel, have become a possibility to replace lead-acid batteries in standby applications [9.20]. Figure 9.12 shows a comparison of the acquisition cost of lead-acid batteries and fuel cells for a 10-year period in a standby power application. The rising cost of batteries is based upon an assumption of replacement at 36- to 60-month intervals [9.21].

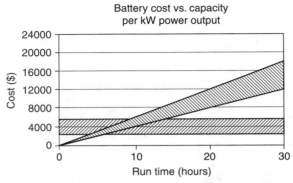

Figure 9.12 Battery cost versus capacity per kW power output. Acquisitions cost comparison for fuel cells and lead-acid batteries in standby power applications. Battery cost-diagonal shading. Fuel cell cost [9.21].
[© 2004, IEEE, reprinted with permission]

Table 9.1 shows the major types of fuel cells considered for standby and alternative electric power use. For applications that require frequent and rapid start-ups, and where hydrogen and air are the available reactants, a polymer-electrolyte membrane fuel cell (PEMFC) is the obvious

TABLE 9.1   Major Types of Fuel Cells. (Advantages vs. Disadvantages) [9.22]

Electrolyte	Operating temp. (°C)	Advantages	Disadvantages
Polymer-electrolyte membrane fuel cell (PEMFC)	60–100	▪ Highest power density ▪ Reduced corrosion and electrolyte-management problems ▪ Rapid start-up time	▪ Relatively expensive catalysts required ▪ High sensitivity to fuel impurities
Alkaline fuel cell (AFC)	90–100	▪ High power density ▪ Demonstrated in space applications	▪ High sensitivity to fuel impurities ▪ Intolerant to CO_2
Phosphoric acid fuel cell (PAFC)	175–200	▪ High quality waste heat (for cogeneration applications) ▪ Demonstrated long life	▪ Relatively expensive catalysts required ▪ Relatively low power density
Molten carbonate fuel cell (MCFC)	600–1000	▪ High quality waste heat ▪ Inexpensive catalysts ▪ Tolerant to fuel impurities	▪ High temperature enhances corrosion and breakdown of all cell components ▪ Relatively low power density
Solid oxide fuel cell (SOFC)	600–1000	▪ High quality waste heat ▪ Inexpensive catalysts ▪ Tolerant to fuel impurities ▪ Solid electrolyte	▪ High temperature enhances corrosion and breakdown of all cell components ▪ Sealing of stacks

[© 2004, IEEE, reprinted with permission]

choice. PEMFC fuel cells also have the highest power density of all of the types in Table 9.1 [9.22]. Fuel cells utilizing hydrogen as fuel can operate for relatively long periods of time—for example, hours—or for short periods, in standby service, in the nature of engine-generator sets or batteries.

Applications.   Specific applications include the following:

- Space (used on Gemini, Apollo, and space shuttle missions)
- As a UPS, which requires instant availability of power when utility service fails. The fuel cell by itself requires heating to start up. UTC Fuel Cells show a 5-kW UPS to supply 48 V dc for telecom applications, which uses ultracapacitors to supply the energy during the start up of the fuel cell system.
- The telecom industry is considering fuel cells as an alternative to VRLA batteries for sites requiring 1 to 3 kW for up to eight hours [9.20–22].
- Nakamoto, et al. show a 4.5-kW fuel cell system to produce 300 V ac power, Figures 9-13 and 9-14 [9.23].
- Utility industry applications include a study by Sedghisigarchi and Feliachi of 1.5-MW fuel cell systems and gas-turbine generators operating on a common bus in a 9-MW system [9.24].

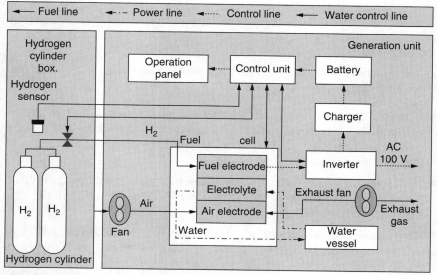

Figure 9.13   Block diagram of 4.5-kW fuel cell UPS [9.23].
[© 2000, IEEE, reprinted with permission]

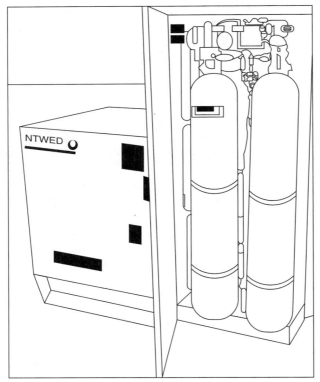

Figure 9.14   External appearance of 4.5-kW fuel cell UPS [9.23]. [© 2000, IEEE, reprinted with permission]

Ultracapacitors

Ultracapacitors can substitute for batteries in low-power UPS. Their features include

- **Construction:** These capacitors utilize electrodes of highly porous carbon to achieve large values of capacitance per unit weight. Zorpetta quotes surface area of 1500 m^2/g; a typical electrode of 250 g would have an area of 375,000 m^2 [9.25].

- **Ratings:** Maxwell Technologies offers ultracapacitors ranging from 5 to 10 F to cylindrical 2700 F, rated 2.5 V DC per cell [9.26]. Storage amounts to 3 or 4 Wh/kg [9.25].

- **Applications:** One manufacturer offers UPS modules rated 1.6 and 2.3 kW, replacing batteries, employing 2300 F of ultracapacitors [9.27]. M. L. Perry describes a 5-kW fuel cell standby power unit in which three banks of ultracapacitors provide the energy while the fuel cells are coming on line [9.22]. Usage of ultracapacitors to supply peak power loads over average power is described by Maher [9.26].

- **Problems:** Because of the low voltage rating of ultracapacitor cells—for example, 2.5 V DC—the cells must be connected in series for a typical 48 V DC application, with the usual problems of cell current and voltage division during charging [9.26]. Cost of $9500/kWh is five times that of lead-acid batteries [9.26].

Summary

The uninterruptible power system is necessary to provide reliable power to critical loads that cannot tolerate interruption. The most widely used equipment is the battery inverter UPS module that is available in power ratings from 100 W to 1000 kW (and higher) at all commercial input and output voltage and phase configurations. Energy storage is usually provided by flooded or VLRA-type batteries for full-load operating times in the range of minutes. Recently developed energy-storage alternatives include flywheels and fuel cells.

References

[9.1] A. Kusko, *Emergency Standby Power Systems*, McGraw-Hill, New York 1989.
[9.2] B. D. Bedford and R. G. Hoft, *Principles of Inverter Circuits*, John Wiley, New York 1964.
[9.3] J. L. Fink, J. F. Johnston, and F. C. Krings, "The Application of Static Inverters for Essential Loads," IEEE Transactions on Power Apparatus and Systems, December 1963, pp. 1068–1072.
[9.4] A. Kusko and F. C. Gilmore, "Concept of a Modular Static Uninterruptible Power System," Conf. Record, Second IEEE IGA Annual Meeting, Pittsburgh, October 1967, pp. 147–153.
[9.5] EC&M, February 2006, p. 58.
[9.6] APC, "Solutions," Winter/Spring 2005, p. 25.
[9.7] Pentadyne Power Corp., Chatsworth, CA, Brochure.
[9.8] A. Kusko and J. DeDad, "Short-Term, Long-Term, Energy Storage Methods for Standby. Electric Power Systems," Conference Record of the 2005 IEEE Ind. Appl. Conf 40th Annual Meeting, Hong Kong, October 2–6, 2005, pp. 2672–2678.
[9.9] C. Robillard, A. Vallee, and H. Wilkinson, "The Impact of Lithium-Metal-Polymer Battery Characteristics of Telecom Power System Design," INTELEC 2004, pp. 25–31.
[9.10] I. Kiyokawa, K. Niida, T. Tsujikawa, and T. Motozu, "Integrated VRLA-Battery. Management System," INTELEC 2000, pp. 703–706.
[9.11] S. S. Misra, "VRLA Battery Development and Reliability Considerations," INTELEC 2004, pp. 296–300.
[9.12] M. R. Cosley and M. P. Garcia, "Battery Thermal Management System," INTELEC 2004, pp. 38–45.
[9.13] R. S. Weissbach, G. G. Karady, and R. G. Farmer, "A Combined Uninterruptible Power Supply and Dynamic Voltage Compensator Using a Flywheel Energy Storage System," IEEE Trans. on Power Delivery, vol. 16, no. 2, April 2001, pp. 265–270.
[9.14] G. Reiner and N. Wehlau, "Concept of a 50 MW/650 MJ Power Source Based on Industry-Established MDS Flywheel Units," Pulsed Power Plasma Science, 2001, vol. 1, pp. 17–22, June 2001.
[9.15] R. G. Lawrence, K. L. Craven, and G. D. Nichols, "Flywheel UPS," IEEE IAS Magazine, May/June 2003, pp. 44–50.

[9.16] I. Takahashi, Y. Okita, and I. Andoh, "Development of Long Life Three Phase Uninterruptible Power Supply Using Flywheel Energy Storage Unit," Proceedings of the 1996 International Conference on Power Electronics, Drives, and Energy Systems for Industrial Growth, 1996, vol. 1, January 8–11, 1996, pp. 559–564.

[9.17] T. Richard, R. Belhomme, N. Buchheit, and F. Gorgette, "Power Quality Improvement Case Study of the Connection of Four 1.6 MVA Flywheel Dynamic UPS Systems to a Medium Voltage Distribution Network," Trans. and Dist. Conference and Exposition, 2001 IEEE/PES, vol. 1, October 28–November 2, 2001, pp. 253–258.

[9.18] D. J. Carnovale, A. Christe, and T. M. Blooming, "Price and Performance Considerations for Backup Power and Ride-Through Solutions," Power Systems World 2004 Conference, November 16–18, 2004.

[9.19] J. R. Sears, "TEX: The Next Generation of Energy Storage Technology," Power Systems World 2004 Conference, November 16–18, 2004.

[9.20] R. Billings and S. Saathoff, "Fuel Cells as Backup Power for Digital Loop Carrier System," INTELEC 2004, pp. 88–91.

[9.21] G. M. Smith and R. P. Spencer, "Enabling Fuel Cells for Standby Power-Chemical Hydride Fueling Technology," INTELEC 2004, pp. 65–72.

[9.22] M. L. Perry and S. Kotso, "A Back-Up Power Solution with No Batteries," INTELEC 2004, pp. 210–217.

[9.23] Y. Nakamoto, N. Kato, T. Hasegawa, T. Aoki, and S. Muroyama, "4.5-kW Fuel Cell System Based on PEFCs," INTELEC 2000, pp. 406–410.

[9.24] K. Sedghisigarchi and A. Feliachi, "Dynamic and Transient Analysis of Power Distribution Systems with Fuel Cells—Part I, II: Fuel Cell Dynamic Model," IEEE Trans. on Energy Conversion, vol. 19, no. 2, June 2004, pp. 423–434.

[9.25] G. Zorpette, "Super Charged," IEEE Spectrum, January 2005, pp. 32–37.

[9.26] B. Maher, "High Reliability Backup for Telecommunications Using Ultracapacitors," INTELEC 2004, pp. 623–625.

[9.27] "Ultracapacitor-Based UPSs Eliminate Batteries," Electronic Products, January 2005, pp. 25–27.

Dynamic Voltage Compensators

*Dynamic voltage compensators for voltage sags and surges
represent a simpler and less costly means for achieving
acceptable power quality than battery-powered UPS.*

Introduction

Conventional utility power-system equipment has long utilized transformers with automatic under-load tap changers to compensate for deviations in line voltage above and below a desired reference level. The circuit for a transformer with a tap changer is shown in Figure 10.1. The controller selects a voltage from the secondary winding and adds or subtracts it from the secondary voltage through a series transformer to produce the desired load voltage at terminal X1 [10.1]. The operation of the mechanical tap changer is too slow to compensate for rapid deviations in line voltage that affect voltage-sensitive equipment. By comparison, the dynamic voltage compensator utilizes power-electronic devices to switch and compensate for line-voltage sags within a half-cycle time to meet the power-quality requirements of computers and other voltage-sensitive equipment.

The operation of the dynamic voltage compensator requires two steps. In the first step, when the source voltage is within an acceptable band, the compensator utilizes a by-pass switch to connect the source directly to the load. In the second step, when the source voltage sags outside of prescribed limits, the compensator injects a correction voltage utilizing power from the source or internal capacitors. The compensator is less costly and more efficient than a battery-powered UPS that requires stored energy in batteries and is usually online continuously supplying the load.

The size, rating, and cost of the components of the compensator depend upon the maximum voltage and time duration of the deviation to be compensated. For sags down to zero voltage lasting up to 0.2 s,

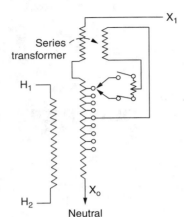

Figure 10.1 Transformer-type tap changer for regulating secondary voltage [10.1].

the compensator is smaller and less costly than a UPS. However, it cannot handle the long-time voltage deviations and total outages that a battery-powered UPS can.

Principle of Operation

An elementary form of dynamic compensator is shown in Figure 10.2 [10.2]. In the circuit of Figure 10.2, under normal conditions, the load is supplied through two back-to-back thyristors acting as a by-pass switch. Meanwhile, the voltage-doubler diode rectifier maintains the dc-link capacitors fully charged. When a voltage deviation—for example, sag—is detected, the controller opens the by-pass switch, turns on the IGBT inverter, and injects the component of missing source voltage using the stored energy in the dc-link capacitors.

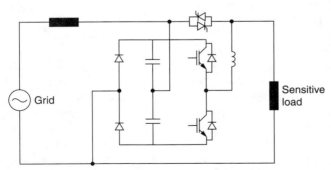

Figure 10.2 Dynamic voltage compensator. Single phase. Parallel injection of voltage correction [10.2].
[© 2005, IEEE, reprinted with permission]

When the voltage compensator utilizes a series transformer to add a voltage component to the source voltage during a voltage sag, the dynamic voltage compensator can be built in one of two forms [10.3]:

- As a shunt converter (rectifier) on the line side of the series transformer (as shown in Figure 10.3a)
- As a shunt converter (rectifier) on the load side of the series transformer (as shown in Figure 10.3b)

In the circuits of Figures 10-3a and 10-3b, the shunt converter is a rectifier that maintains the required voltage on the dc-bus capacitor. The series converter is a PWM inverter that produces the compensating voltage for correction of the source voltage during a voltage sag.

In the circuits of Figure 10.3, under normal conditions, the load is supplied through the line-side winding of the series transformer. The upper IGBTs in the series converter are turned "on" so as to short circuit the converter-side winding of the series transformer and reflect a low impedance to the line-side winding. When a voltage deviation is detected, the

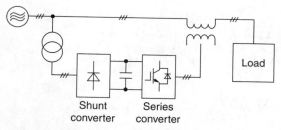

Figure 10.3a Dynamic voltage compensator. Series injection of correction voltage. Shunt converter (rectifier) on the source side [10.3].
[© 2005, IEEE, reprinted with permission]

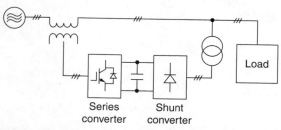

Figure 10.3b Dynamic voltage compensator. Series injection of correction voltage. Shunt converter (rectifier) on the load side [10.3].
[© 2005, IEEE, reprinted with permission]

series converter is activated; it injects the missing component of the source voltage waveform. In the circuit shown in Figure 10-3b, the energy comes from the shunt transformer on the load side, not from the dc-link capacitors. However, the line current increases to maintain the power to the load constant. As a result, the time duration of the compensation does not depend upon the energy stored in the capacitors but on the short-time thermal capabilities of the components.

Figure 10.4 shows an experimental result of dynamic voltage compensation for a single-phase voltage sag with a depth of 50 percent using the circuit of Figure 10.3b [10.3]. The voltage sag lasts 110 ms. The waveform v_C is the series transformer voltage, which is added to the source voltage v_S to produce the load voltage v_L.

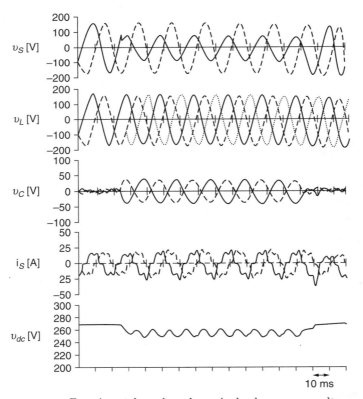

Figure 10.4 Experimental result under a single-phase source voltage sag of 50% for 110 ms = source voltage; v_L = load voltage; v_C = correction voltage; i_s = source current; v_{dc} = capacitor voltage [10.3]. [© 2005, IEEE, reprinted with permission]

Operation on ITIC curve

With a relatively limited range of operation, the dynamic voltage compensator (without energy storage) can protect loads for a majority of voltage sags that fall outside the "acceptable" region of the CBEMA or ITIC curves. The operation of a commercial dynamic voltage compensator is superimposed on the ITIC and CBEMA curves in Figure 10.5 [10.4]. The scatter plot of sags below the ITIC curve (dashed lines) fall into the region covered by the compensator. The commercial compensator utilizes the parallel circuit of Figure 10.2 for single-phase operation and the series circuit of Figure 10.3 for higher power three-phase operation up to 500 kVA [10.4]. This particular compensator can provide a 100 percent boost of the source voltage for a time up to 0.2 s (12.4 cycles), and a 50 percent boost up to 2 s (124 cycles). The compensator utilizes the energy supplied from the line or stored in the dc-link capacitors. For a 50-percent boost, the capacitors require 60 s to recharge before the compensator can handle another sag of time duration up to 2 s.

Longer times of operation than that shown in Figure 10.5 require energy storage—for example, batteries—in the dc link. At some sag time duration, a battery-powered UPS provides a better solution for the power-quality problem. The UPS can handle both short time and extended time outages. A comparison of the dynamic voltage compensator and the UPS is given in reference [10.5], and also shown in the table of Figure 10.6 [10.4].

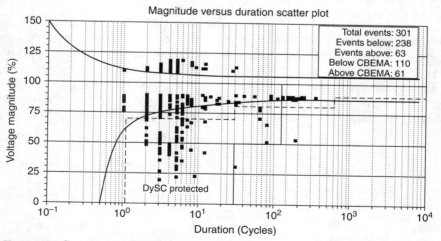

Figure 10.5 Scatter plot of PQ events at one industrial site over 2.3 years, overlaid with the CBEMA curve (solid thin lines), the ITIC curve (dashed lines), and the single-phase compensator (DySC) protection regime (thick lines) [10.4].
[© 2001, IEEE, reprinted with permission]

Type of event	% of total PQ events	Spike suppressor	Voltage regulator	Transformerless DySC	UPS systems
Spikes & surges	5%	Solves	Solves	Solves	Solves
Sag to 80%	35%	No	Solves	Solves	Solves
Sag from 50–80%	45%	No	No	Solves	Solves
Interruption 0–0.15 s	7%	No	No	Solves	Solves
Interruption 0.15–500 s	4%	No	No	No	Solves
>500 sec outage	4%	No	No	No	No
Total PQ events solved	100%	5%	40%	92%	96%
kVA range		1–1000 kVA	1–200 kVA	1–2000 kVA	0.2–1000 kVA
kVA/lb		1–10	0.02–0.03	0.2–1.0	0.01–0.02
kVA/ft^3		500	1–2	10–50	0.3–1

Figure 10.6 Table 10.1 protection capability for various types of power conditioning equipment for typical percentages of power-quality events [10.4].
[© 2001, IEEE, reprinted with permission]

Detection of disturbance and control

The control circuit for the dynamic voltage compensator performs the following three functions:

- Detects the voltage disturbance. Determines whether it exceeds the prescribed limits so that compensation of the source voltage is required
- Opens the by-pass switch between the source and the load
- Initiates the operation of the converter that supplies the missing portion of the source-voltage waveform

Several control circuits have been described in the literature for dynamic voltage compensators. The block diagram of one control circuit is shown in Figure 10.7 [10.3]. The blocks are typical of other published control circuits to perform the functions previously described.

In Figure 10.7, the three-phase line voltage v_s is transformed into two components: direct-axis voltage v_d and quadrature-axis voltage v_q.

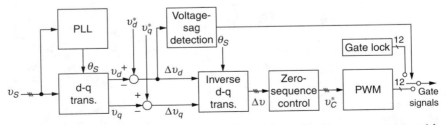

Figure 10.7 Block diagram of the control circuit for a dynamic voltage compensator with a series injection of correction [10.3].
[© 2005, IEEE, reprinted with permission]

For balanced line voltages, $v_d = 1$ and $v_q =$ zero. The phase-locked loop circuit PLL detects the line voltage v_s, operates in synchronism with the line voltage, and calculates the phase angle θ_s, which is used in the transformation of v_s.

The voltages v_d and v_q are compared with reference voltages v_d^* and v_q^* to yield the error voltages Δv_d and Δv_q, which represent the deviation of the source voltage from the referenced waveform and amplitude. The voltage sag detection circuit responds to a value of Δv_d that exceeds a preset value—for example, 2 percent. The circuit opens the by-pass switch (Gate Lock) and feeds gate signals to the compensating converter. The error voltages Δv_d and Δv_q undergo an inverse d-q transformation to produce a three-phase set of deviation voltages (v, which become the three-phase reference voltages v_c^* for the converter that provides the components of compensation voltage.

Commercial equipment

An example of commercial equipment is shown in Figure 10.8 [10.6]. The voltage compensators range from 250 VA single-phase to 333 kVA three-phase. The three-phase dynamic voltage compensator is rated 480 V, 400 A, 333 kVA. The particular unit can compensate for the following sags for the time duration:

- Single line to zero voltage remaining, 2 s
- Two phases, to 30 percent voltage remaining, 2 s

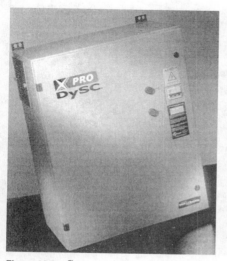

Figure 10.8 Commercial dynamic voltage compensators, 250 VA to 333 kVA [10.6]. [Courtesy SoftSwitching Technologies]

- Three phases, to 50 percent voltage remaining, 2 s
- Three phases, to zero voltage remaining, 3 cycles standard, 12 cycles with extended outage provision

Summary

The dynamic voltage compensator typically corrects line voltage sags down to zero percent remaining for a time of 12 cycles, and to 50 percent remaining for up to 2 s. The compensator does not require any stored energy, such as batteries. Therefore, the compensator is a less costly equipment than a battery-powered UPS. However, the compensator cannot supply power to critical loads for outages longer than 2 s, while a battery-powered UPS is only limited in time by the ampere-hour capacity of its batteries.

References

[10.1] D. G. Fink and H. W. Beaty, *Standard Handbook for Electrical Engineers*, 14th edition, McGraw-Hill, 2000, Figure 10.28.
[10.2] S. M. Silva, S. E. da Silveira, A. de Souza Reis, and B. J. Cardosa Filho, "Analysis of a Dynamic Voltage Compensator with Reduced Switch-Count and Absence of Energy Storage Systems," IEEE Trans. Ind. Appl., vol. 41, no. 5, September 10, 2005, pp. 1255–1262.
[10.3] T. Jimichi, H. Fujita, and H. Akagi, "Design and Experimentation of a Dynamic Voltage Restorer Capable of Significantly Reducing an Energy-Storage Element," Conference Record, 2005 40th IAS Annual Meeting, pp. 896–903.
[10.4] W. E. Brunsickle, R. S. Schneider, G. A. Luckjiff, D. M. Divan, and M. F. McGranaghan, "Dynamic Sag Correctors: Cost-Effective Industrial Power Line Conditioning," IEEE Trans. Ind. Appl., vol. 37, no. 1, January/February 2001, pp. 212–217.
[10.5] A. Kusko and N. Medora, "Economic and Technical Comparison of Dynamic Voltage Compensators with Uninterruptible Power Supplies," Power Quality Conference, October 24–26, 2006.
[10.6] SoftSwitching, "Technical Bulletin," Middleton, WI.

Power Quality Events

Previous chapters have described power-quality deficiencies such as line-voltage sags, voltage interruption, and the effects of harmonic currents, as well as the remedies for these conditions. In this chapter, we will describe the sensitivities and reactions of equipment subjected to these deficiencies.

Equipment affected by power-quality deficiencies and described in this chapter include the following:

- *Personal computers*
- *Controllers*
- *AC and DC contactors*

(Electric motor drive equipment is covered in Chapter 12.)

Introduction

The purpose for analyzing the electrical system and the response of critical equipment to power-quality deficiencies is to minimize the misoperation of each piece of equipment and the system in which it is located. The following two comprehensive methods of analysis can be used.

Method 1

C. P. Gupta et al. described a theoretical method incorporating the following steps [11.1]:

1. Prepare a network model of the electrical system, including the protective equipment and the connection of the critical load equipment.

2. Apply faults to the network lines and buses in accordance with average industry fault data.

3. Compare the resulting voltages at the connections to equipment with the known sensitivities of the equipment.

4. Calculate the number and duration of the interruptions per year of each piece of critical equipment.

5. Make decisions on correction measures.

Method 2

1. Review all critical load at a facility, ranging from personal computers to complete industrial processes or data processing systems.

2. Estimate the cost of an interruption to each piece of equipment.

3. Carry out prevention and correction measures at a cost commensurate with that of an interruption.

In the real world, the design of an electrical system to supply critical load is based on many factors: required availability, cost, space, record of utility service, and others. The limit in system design occurs when the complexity of the correction measure overrides the predicted reliability and availability of the system.

Personal Computers

A personal computer (PC) is a general-purpose computing device designed to be operated by one person at a time. PCs can exert real-time control of external devices, operate online control of communications, or be part of process-control applications. The malfunction of PCs incorporated in a real-time system because of voltage disturbances produces bigger consequences than the malfunction of the PC used offline [11.2].

A first measure to match the capabilities of PCs to line-voltage disturbances was the CBEMA curve shown in Figure 11.1, from IEEE Std 446-1987 [11.3 and 11.4]. The curve defines the tolerance level of "automatic data processing" equipment to voltage sags, swells, and short interruptions. The curve was updated in 1995 to the ITIC curve for "information technology equipment" for 120-V/60-Hz single-phase service. A similar curve, SEMI F47, was proposed for "semi-conductor processing" equipment. The three curves are shown in Figure 11.2 [11.2].

An example of measured voltage disturbances at a customer's site is shown in Figure 11.3 [11.5]. Here, the concern was damage to the PC from voltage surges. PCs can withstand surges of up to 3 kV.

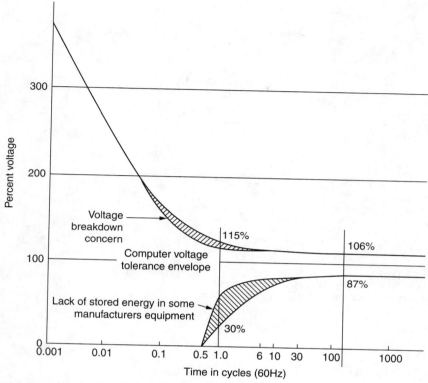

Figure 11.1 Early CBEMA curve. Typical computer voltage tolerance envelope. Source, IEEE Std 446-1987 [11.3].
[© 1987, IEEE, reprinted with permission]

Power-quality characteristics

Personal computers both produce disturbances to the power system and are affected by voltage disturbances from the system. Each PC incorporates a switch-mode power supply to convert power from the AC supply line to low-level DC voltages for the internal circuits. The older PCs utilized an input diode bridge with a DC capacitor filter. Newer PCs are designed with an input PWM circuit that shapes the line current to a sinusoidal waveform in phase with the line voltage—the so-called unity power factor operation.

The waveform of the line current of older PCs is shown in Figure 11.4 [11.6]. The dominant harmonic is the third. When a group of older PCs is supplied from a three-phase, 120/208-V power panel, the third-harmonic currents return from the panel to the source in the neutral conductor. As a result, facilities have to be wired with oversize or "double" neutrals. The unity power factor circuit in the new PCs reduces the requirement.

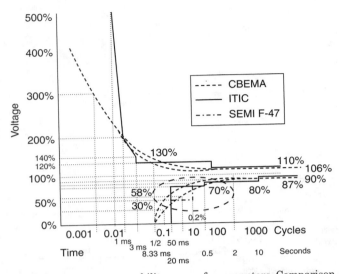

Figure 11.2 Power acceptability curves for computers. Comparison of CBEMA ITIC and SEMI F47 curves [11.2].
[© 2005, IEEE, reprinted with permission]

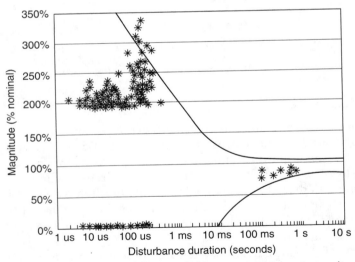

Figure 11.3 A voltage disturbance profile at a commercial customer site. Surges and sags during 2074 hours [11.5].
[© 1998, IEEE, reprinted with permission]

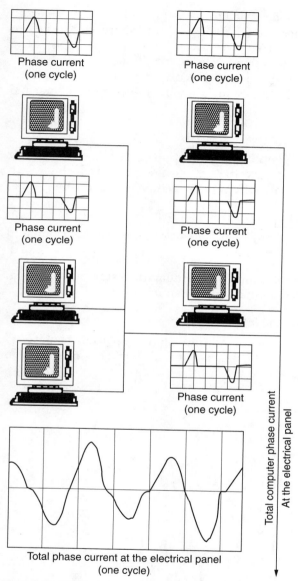

Figure 11.4 Total phase current to a network of personal computers [11.6].
[© 1997, IEEE, reprinted with permission]

Personal computers are designed to withstand line voltage sags and surges in accordance with the CBEMA curve of Figure 11.1 or better. Actual test results of sensitivity are given later in this chapter. The response of an older PC to a four-cycle interruption of line voltage is shown in Figure 11.5 [11.5]. The inrush current to recharge the filter capacitor upon voltage restoration is about 300 percent of normal current.

Modes of malfunction

The modes of personal computer malfunction under line voltage sag occur as the DC filter capacitor voltage of the power supply declines with time. The ensuing software malfunctions include the following [11.2]:

- Lockup, interruption, or corruption of read/write operations (blue screen)
- Blocking of the operating system, lack of response to any command from the keyboard (frozen screen)

Hardware malfunction is identified by automatic restarting/rebooting, or a permanent black screen, making a manual restart necessary.

Sensitivity to voltage sags and interruptions

Voltage sag tests on personal computers show that their sensitivity to voltage disturbances follows a rectangular curve, as shown in

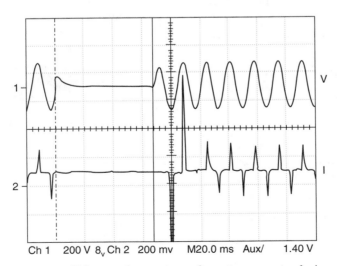

Ch 1   200 V 8ᵥ Ch 2   200 mv   M20.0 ms   Aux/   1.40 V

Figure 11.5 Voltage and current waveforms to a computer during and following a momentary outage [11.5].
[© 1998, IEEE, reprinted with permission]

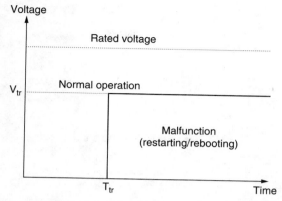

Figure 11.6 Rectangular voltage-time tolerance curve of a personal computer [11.2].
[© 2005, IEEE, reprinted with permission]

Figure 11.6 [11.2]. Depending upon the PC, the voltage V_{tr} ranges from a low of 30 to 40 percent to a high of 80 to 90 percent. The test results for a specific PC are shown in Figure 11.7 as a function of the malfunction of the PC. For restart/rebooting, the interruption can last 380 ms, or 23 cycles. A longer voltage sag can extend down to 20 percent of the rated voltage [11.2].

Tests of the effects of typical utility power disturbances were made on a PC LAN test system shown in Figure 11.8 [11.5]. Disturbance signals were introduced to the line supplying the LAN. The response of the PCs is shown in Figure 11.9 [11.5].

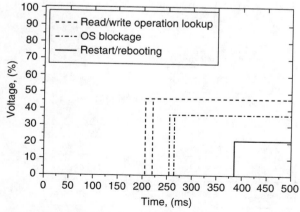

Figure 11.7 Voltage-time tolerance curve for personal computer as a function of failure mode [11.2].
[© 2005, IEEE, reprinted with permission]

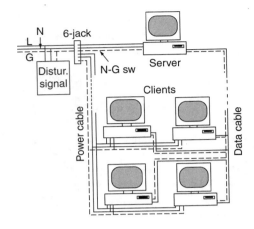

Figure 11.8 Test systems for voltage disturbance into a personal computer LAN [11.5].
[© 1998, IEEE, reprinted with permission]

Correction measures

For continuous operation of personal computers or other sensitive systems when the line voltage interruptions last approximately 0.5 s or longer, the only solution is a UPS [11.4].

Controllers. A controller is defined as a device or group of devices that serves to govern, in some predetermined manner, the electric power delivered to the apparatus to which it is connected [11.7]. Controllers can operate in hydraulic, mechanical, and other systems, as well as electrical. In all cases, the controller usually receives its operating power, and information, from the utility supply line. Examples of controllers are the speed regulator of a motor-drive system, the temperature controller of an industrial furnace, and the voltage regulator of a controlled rectifier. A block diagram of a process controller is shown in Figure 11.10 [11.8].

Disturbance test	Problems monitored				N-G bond
	PC Look-up	Network slowdown	File corrup.	Monitor	
Momnt. inter.	No	Probable*	No	Brief flicker	No effect
Cap. switching	No	No	No		No effect
Lighting surge	No	No	No		No effect
Local appl.	No	No	No		No effect
EFT	No	Yes	Probable*	Flicker & buffer corr.	No effect
RFI	Yes	Yes	Probable*	Flicker & buffer corr.	No effect

* See the corresponding test results for more detail.

Figure 11.9 Table 1. Summary of test results on LAN test systems [11.5].
[© 1998, IEEE, reprinted with permission]

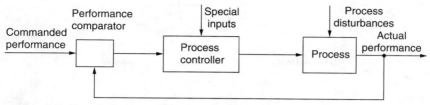

Figure 11.10 Functional block diagram of a controller [11.8].

Inputs to controllers. The inputs to controllers include such items as the following:

- **Manual:** switches, push buttons
- **Sensors and References:** voltages, position, speed, temperature
- **Data Links:** information, command
- **AC Voltage:** operating power, frequency, and phase information

Output from controllers. The outputs, which operate and control power devices, include the following:

- Contactors, motors, furnaces, elevators
- Power semiconductors, thyristors, GTOs, MOSFETS

Design. The controller is designed to provide output functions in response to the input signals. The design can be based on one or more of the following:

- **Relay Logic:** Using AC or DC relays to execute on-off operations
- **Digital Semiconductor Logic:** Using individual packages or a PLC, as shown in Figure 11.11 [11.9]
- **Digital Computer:** personal computers or larger, programmed for the controller function

In all cases, the operating power comes from the utility supply line. For controllers that deliver gating signals to power semiconductors, the supply line provides voltage, frequency, and phase information.

Disturbances. When a line-voltage sag occurs, a controller will fail if the percent sag exceeds the sensitivity of the relays or the logic circuits supplied from the internal switch-mode power supply. The sensitivity of the relay is treated in the Section "AC contactors and relays" later in this chapter. The sensitivity of the switch-mode power supplies and logic circuits corresponds to the information in Section "Personal

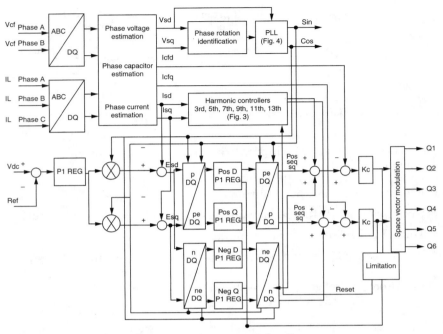

Figure 11.11   Controller utilizing digital components [11.9].
[© 2005, IEEE, reprinted with permission]

Computers". Where the controller requires phase information from the supply line, such as for base drive signals in Figure 11.12, the controller might be more sensitive to voltage sags than the personal computer.

Correction measures

To keep a controller operating in the face of line-voltage sags and interruptions, the following equipment can be used:

- **Constant-Voltage Transformers (CVs):** Used to correct for sags down to 30 percent of rated line voltage for one cycle, and down to 70 percent continuously for a full load, as described in Chapter 8.

- **UPSs:** For all conditions except line-voltage phase information. Assume that the controlled load is operating from the supply line to the UPS so that phase information from the UPS is of no use during deep and long sags or interruptions.

- **Phase-Locked Loop:** Used in the controller to provide phase information.

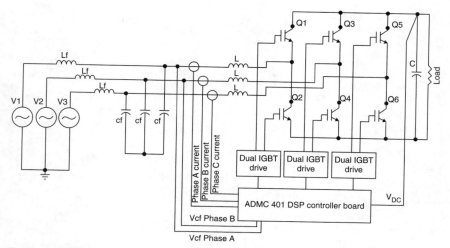

Figure 11.12 Three-phase power factor correction boost converter. Line filter. Controller supplies base drive [11.9].
[© 2005, IEEE, reprinted with permission]

AC Contactors and relays

A contactor is an electromagnetic device in which the current in the operating coil causes the armature to move against a spring force so as to close or open electrical contacts. The contactor is widely used, as a motor starter, to control electric furnaces and heaters, and wherever an electric-power circuit must be operated from a low-power electrical control signal. A relay is a smaller electromagnetic version of the contactor. A typical application is shown in the diagram of Figure 11.13, where the Emergency Off Relay EMO is actuated by the pushbutton "On," and the EMO Relay Contact energizes the main contactor coil [11.10].

Contactors and relays are built for AC or DC operation in a variety of coil voltage and contact ratings. In some cases, such as in "soft" motor starters, the contactor function has been displaced by power-electronics devices such as GTOs. Likewise, the control relay functions have been displaced by PLCs using digital logic. Whether electromagnetic or solid-state, the devices are impacted by line-voltage sags and interruptions. In a study of 33 "tools" used in the semiconductor manufacturing industry, the circuit shown in Figure 11.13 for motor control was the most susceptible to fail from a voltage sag—of the failures, 33 percent for the relay, 14 percent for the contactors [11.10].

Operation

A contactor or relay device "fails" when the position of its armature and contacts do not comply with the signal delivered to it by its control

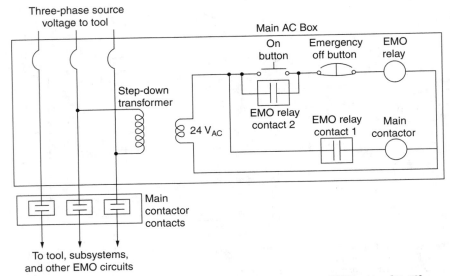

Figure 11.13   Typical main contactor control circuit. Emergency OFF button [11.10].

system. For example, in Figure 11.13, the "On" button and EMO Relay Contact 2 "order" the EMO Relay and the Main Contactor to close their contacts, to start and maintain power to the motor. A sag or interruption in line voltage can cause either or both the relay and contactor to "drop out," open their contacts, and shut down the motor. The contactor or relay has "failed" from the voltage disturbance.

The elementary magnetic circuit for the contactor or relay is shown in Figure 11.14 [11.13]. The device is a DC relay, activated by a DC coil current. The current produces a magnetic field B_m in the air gap. The force in the armature is proportional to B_{m^2}. At a given coil current, the

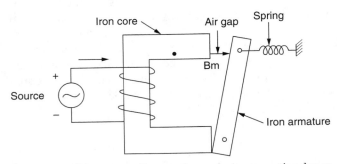

Figure 11.14   Elementary diagram for an electromagnetic relay or contactor [11.13].

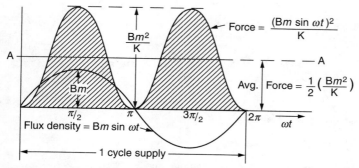

Figure 11.15 Magnetic air-gap flux density. Instantaneous and average force relationships for the AC electromagnetic device [11.11].

force overcomes the spring force, closes the armature and keeps it closed. Depending upon the design of the device, the contacts activated by the armature are either open or closed when the armature is closed. If the voltage source supplying the coil current sags or is interrupted, depending upon the time duration, the spring can pull the armature sufficiently open to reverse the contact positions, and thus the device has failed in its function.

When the device, the relay, or the contactor is activated from an AC source, as shown by the transformer in Figure 11.13, the coil current is alternating. The magnetic field B_m is alternating, as shown in Figure 11.15 [11.11]. The force, proportional to B_{m^2} is unidirectional, reaching a peak twice per cycle. The effect is vibration and noise in the device. The solution is the "shading coil," which is a closed one-turn coil placed in part of the pole face of the device, as shown in Figure 11.16 [11.11]. The magnetic field through the shading coil is delayed each half cycle from the magnetic field in the rest of the pole face. The consequence is a force pattern, as shown in Figure 11.17. The net force is nearly continuous with an average value, as in a DC magnet [11.11].

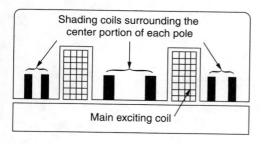

Figure 11.16 Shading coil on pole face of an AC electromagnetic device [11.11].

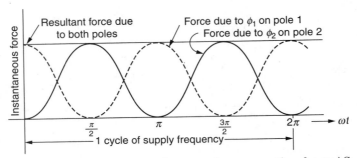

Figure 11.17 Instantaneous and average force versus time for an AC electromagnetic device with a shading coil [11.11].

The Impact of Voltage Disturbance

The device, the relay, or the contactor can be subjected to voltage sag, to a voltage interruption, or both. The theoretical response to the disturbance is shown in Figure 11.18 [11.12]. The device will function normally for voltages down to V_u. For voltage down to zero, on interruption, the device will not fail (drop out) for a time duration T_u. For voltages less than V_u and time durations greater than T_u, the device will fail. The curve of Figure 11.18 is applicable to DC relays or contactors.

The response of an AC relay or contactor depends on the point in the voltage waveform where the voltage disturbance occurs. The waveform in Figure 11.19 shows a voltage sag to 30 percent occurring at the 90° point in the voltage wave. Because the coil of the relay is inductive, the coil current is near zero when the voltage is a maximum at 90°, and the magnetic force is minimum.

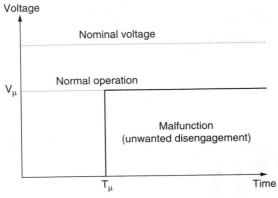

Figure 11.18 Rectangular voltage-time tolerance curve for malfunctions of an AC electromagnetic device [11.12]. [© 2004, IEEE, reprinted with permission].

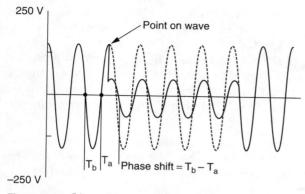

Figure 11.19 Line voltage sag at the 90° point on the waveform of voltage applied to an AC electromagnetic device [11.12] [© 2004, IEEE, reprinted with permission].

The generic voltage tolerance curve for AC relays and contactors is shown in Figure 11.20 [11.12]. The experimentally determined values are: $S1(90°) = 5–10$ ms, $S1(0°) = 80–120$ ms, $VS(0°) = 38–75\%$ of nominal voltage, $VS(90°) = 35–60\%$ of nominal voltage. Obviously, the device is more tolerant of a voltage sag or interruption that is initiated at the 0° point on the voltage wave than the 90° point, for the reason previously stated.

Correction methods

Several methods are available to reduce the impact of voltage sags and interruptions on relays and contactors. Each method has advantages and disadvantages, as follows:

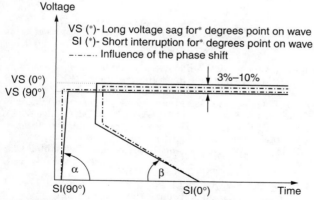

Figure 11.20 Generic voltage-time tolerance curves for an AC electromagnetic device as a function of the point of the voltage wave where the sag occurs [11.12]. [© 2004, IEEE, reprinted with permission]

- **Constant-voltage transformers (CVs)** can be used to supply the control circuits in the specific equipment—for example, the transformer in Figure 11.13. The CV will deliver proper voltage for voltage sags down to 30–70 percent of nominal voltage, depending on the loading. However, an AC contactor requires an inrush of about six times its operating current to close. The CV is load current limited; it will require an oversize rated CV to supply AC contactors, but not smaller control relays.

- **Energy storage devices** incorporating capacitors or batteries can be used to hold in a relay or contactor when the supply voltage sags or is interrupted. However, the opening function of the device is delayed, which might interfere with its purpose. For example, a motor contactor must trip rapidly for a fault in the motor.

- **DC contactors and relays** can be supplied from rectifiers in AC circuits. The coils can be shunted with capacitors to provide a short-time holding current when the source voltage sags or is interrupted. As in the preceding item listed, the delay in the opening function might interfere with its purpose.

Summary

The sensitivities of personal computers, controllers, AC contactors and relays to power quality deficiencies, along with methods of correction, were addressed in this chapter. The same analysis will be applied to induction motors and adjustable speed drives (ASDs) in the next chapter as examples of industrial power equipment.

References

[11.1] C. P. Gupta and J. V. Milanovic, "Probabilistic Assessment of Equipment Trips due to Voltage Sags," IEEE Trans. on Power Delivery, vol. 21, no. 2, April 2006, pp. 711–718.

[11.2] S. Z. Djokic, J. Desmet, G. Vaneme, J. V. Milanovic, and K. Stockman, "Sensitivity of Personal Computers to Voltage Sags and Short Interruptions," IEEE Trans. on Power Del., vol. 20, no.1, January 2005, pp. 375–383.

[11.3] IEEE Std 446-1987, "IEEE Recommended Practice for Emergency and Standby Power Systems for Industrial and Commercial Applications," (IEEE Orange Book).

[11.4] IEEE Std 1250-1995, "IEEE Guide for Service to Equipment Sensitive to Momentary Voltage Disturbances," Art 5.1.1, Computers.

[11.5] M. E. Barab, J. Maclage, A. W. Kelley, and K. Craven, "Effects of Power Disturbances on Computer Systems," IEEE Trans. on Power Del., vol. 14, no. 4, October 1998, pp. 1309–1315.

[11.6] D. O. Koval and C. Carter, "Power Quality Characteristics of Computer Loads," IEEE Trans on Ind. Appl., vol. 33, no. 3, May/June 1997, pp. 613–621.

[11.7] IEEE 100, *The Authoritative Dictionary of IEEE Standard Terms*, seventh edition, 2000, p. 234.

[11.8] J. G. Truxal, *Control Engineers' Handbook*, first edition, McGraw-Hill, New York, 1958, pp. 3–10.

[11.9] l. Mihalache, "A High-Performance DSP Controller for Three-Phase PWM Rectifiers with Low Input Current THD under Unbalanced and Distorted Input Voltage," Conference Record of the 2005 IEEE Industry Application Conference 40th IAS Annual Meeting, pp. 138–144.

[11.10] M. Stephens, D. Johnson, J. Soward, and J. Ammenheuser, "Guide for the Design of Semiconductor Equipment to Meet Voltage Sag Immunity Standards," Technology Transfer #9906376OB-TR, International SEMATECH, December 31, 1999.

[11.11] H. C. Rotors, *Electromagnetic Devices*, John Wiley, New York, 1941.

[11.12] S. Z. Djokic, J. V. Milanovic, and D. S. Kischen, "Sensitivity of AC Coil Contactors to Voltage Sags, Short Interruptions, and Undervoltage Transients," IEEE Trans. on Power Delivery, vol. 19, no. 3, July 2004, pp. 1299–1307.

[11.13] A. E. Fitzgerald, C. Kingsley, Jr., and A. Kusko, *Electric Machinery*, 3rd edition, McGraw-Hill, 1971.

12

Electric Motor Drive Equipment

In this chapter we discuss issues related to the use of electric motor drive equipment, specifically equipment, operation and protection for induction motors.

Electric Motors

Every industrial, commercial, or residential facility utilizes electric motors, ranging from a fractional horsepower motor for a cooling fan to thousands of horsepower in a motor for a plastic extruder. The power for these motors is either supplied from the utility line or from power-electronic inverters in adjustable speed drive (ASD) systems. They are all susceptible to power-quality problems—for example, line-voltage disturbances. In this chapter, we will address the vulnerabilities of both induction motors and ASDs, as well as methods to reduce their vulnerability to these problems.

Induction Motors

Induction motors provide mechanical power to a wide range of domestic, commercial, and industrial loads, ranging from refrigerators to machine tools. These motors operate from single-and three-phase power sources, are controlled by starters, and are usually protected by fuses and thermal overload devices, as shown in Figure 12.1 [12.2]. The motors operate essentially at constant speed unless powered by an inverter in ASDs. As such, the induction motors are subject to all of the voltage dips, interruptions, unbalance, and harmonics that characterize poor power quality. For information on induction motors, see [12.4].

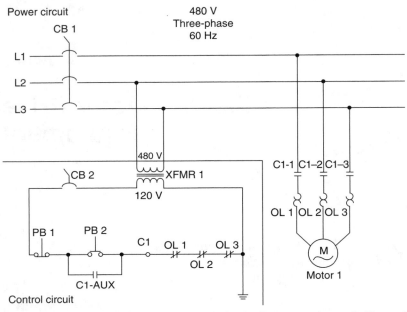

Figure 12.1 Basic induction motor control circuit. Starter contactor C, Thermal overloads OL. Circuit breaker CB [12.2].
[© 2002, IEEE, reprinted with permission]

Operation

The reaction of the induction motor to disturbances is governed by the following factors [12.2, 12.3]:

- The disturbance is a voltage sag, typically to 70 percent rated voltage for up to 10 cycles, or an interruption for up to 1 minute.

- The disturbance either affects all three phases, or one or two phases, resulting in unbalanced voltages.

- The motor was operating either at no load or loads up to full inertia load.

- The motor is disconnected from the line by the disturbance, is reconnected (restarted), or stays connected.

- The motor is limited on the number of restarts [12.4].

Hazards

After the induction motor is subjected to a sag or total interruption of line voltage, the hazards include the following:

1. The motor stops either because a contactor has dropped out, or because the motor current blew the fuses as it attempted to recover

speed or restart, when power was restored. The hazard depends on the inertia and function of the load that the motor was driving.

2. The motor damages itself from high stator current or negative shaft torque when it attempts to recover its speed (particularly under high inertia load) after a short-time sag or interruption of line voltage [12.3].

Phenomena

In the normal operation of the induction motor, the rotating air-gap magnetic field links both the stator windings and the rotor windings (squirrel-cage bars). When the stator voltage sags or is interrupted, two phenomena occur: the rotor slows down (depending on the load inertia) and the air-gap magnetic field declines in amplitude but is supported by the rotor currents.

When the stator voltage is restored, depending upon the outage time, the stator-produced magnetic field and the declining rotor air-gap field can be out of phase. The result is nearly double the line voltage acting on the leakage reactance, resulting in a large transient stator current and negative shaft torque. The oscillograms for a 10-hp motor with an inertia load of 4.28 per unit (pu) of the rotor inertia and a 7.3 cycle interruption are shown in Figure 12.2 [12.3]. The peak current is 1.67 pu of

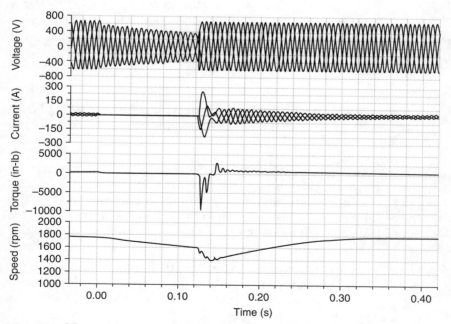

Figure 12.2   Motor voltage, current, shaft torque, shaft speed versus time for 7.3 cycle voltage interruption for a 10-hp motor. [12.3].
[© 2006, IEEE, reprinted with permission]

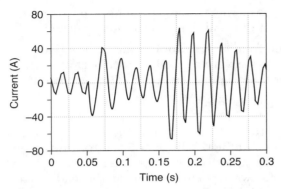

Figure 12.3   7.5-hp induction motor. Current during voltage sag to 30 percent of 5.5 cycles and after voltage is restored [12.2].
[© 2002, IEEE, reprinted with permission]

the starting current; the peak negative torque is 1.71 pu of the starting torque.

The impact on the induction motor of line-voltage sag is similar for an interruption of the same time duration where the motor voltage is restored. Figure 12.3 shows the stator current of a 7.5-hp induction motor subjected to a voltage sag to 30 percent for 5.5 cycles. The motor was loaded only with the inertia of an eddy-current brake. The current at the restoration at the end of the voltage sag is about 4.5 times rated current, of the order of the line starting current [12.2].

Protection

Measures are available to protect the induction motor and its load from the effects of line voltage sag and interruption. They include the following:

1. Drop out of the motor starting contactor can be delayed or prevented with a contactor ride through device, CRD1, as shown in Figure 12.4 [12.3]. The device utilizes some type of energy storage to keep contactor C1 closed. This protection is most suitable for single-motor, single-load, applications—for example, pumps and fans, as compared to multimotor process systems.

2. Operate the induction motor in an adjustable speed drive where the dc-link capacitor voltage is maintained during a line-voltage sag or interruption by an auxiliary source from the kinetic energy of the load, as shown in Figure 12.5.

3. Control the induction motor with a microprocessor-based motor management relay. The relay will regulate restarts after voltage

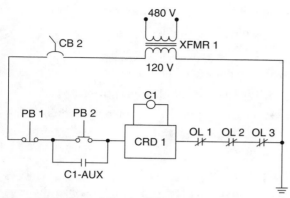

Figure 12.4 Control circuit of Figure 12.1 with voltage sag contactor ride-through device CRD1 [12.2].
[© 2002, IEEE, reprinted with permission]

interruptions based on the thermal state of the motor, as compared with a count of starts per hour. The relay is suitable for large motors where the numbers of starts is limited to prevent overheating [12.5].

Adjustable Speed Drives

An adjustable speed drive (ASD) consists of a rectifier, an induction motor, and a controller. The inverter provides power to the induction motor at a frequency typically from near zero to 120 Hz, so that the induction motor speed is adjustable from near zero to twice rated. The inverter and motor voltage is controlled as the frequency is changed on a constant volts-per-Hertz basis to maintain constant air-gap flux density in the motor in so-called constant torque operation.

A simplified diagram of an ASD supplied by a three-phase line is shown in Figure 12.5 [12.7]. The essential parts from left to right are the following:

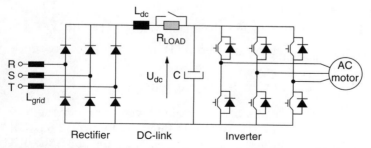

Figure 12.5 A typical ASD with diode rectifier bridge, line, and DC-link reactors and braking resistor [12.7].
[© 2004, IEEE, reprinted with permission]

- Line reactors, L_{grid}, to reduce line-current harmonics
- Diode rectifier for supplying power to the dc-link
- Filter inductor, L_{dc}
- Braking resistor, R_{load}, to absorb energy when the motor is decelerating
- A dc-link capacitor, C, to filter the output of the rectifier
- Voltage source PWM inverter to supply three-phase sinusoidal voltage AC power to the motor
- Three-phase AC induction motor

An alternative design for large ASDs uses a current-source inverter. The inverter delivers a square-wave current to the motor. Both the rectifier and inverter employ thyristors or GTOs [12.7].

Application

ASDs represent one of the largest groups of three-phase loads in industry that both impact their supply systems and also are affected by voltage disturbances imposed by the supply system. ASDs are built in the range from 1–10,000 hp for applications where the speed of the induction motor must be controlled. Applications include pumps, fans, machine tools, and the soft starting of large motors. Drives from 1 to 500 hp typically employ PWM voltage-source inverters, while drives from 300 to 1000 hp and larger typically employ current-source inverters [12.6].

The rectifier part of the ASD produces harmonics in the line currents. For example, the diode rectifier in Figure 12.5 produces fifth, seventh, and higher-order harmonics, as described in Chapter 5. The effect on electrical supply systems, as harmonic voltages at the point of common coupling (PCC) to other loads, can be reduced using the following measures [12.6]:

- Reduce the line impedance from utility source to PCC.
- Install active and passive filters electrically close to the ASD.
- Utilize phase multiplication in the ASD rectifiers—for instance, 12 pulse.
- Utilize a PWM rectifier to shape line currents.

For larger ASDs, 12-pulse rectifiers are employed to reduce the line-current harmonic level. Such a converter is shown in Figure 12.6 [12.6]. The input transformer in large-hp drives can serve to step down the utility or plant primary voltage and to provide the phase-shifted delta and wye secondary voltages for the two rectifier bridges. The individual and

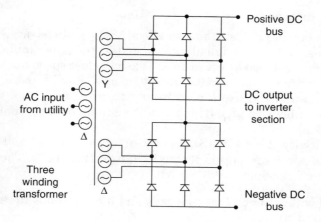

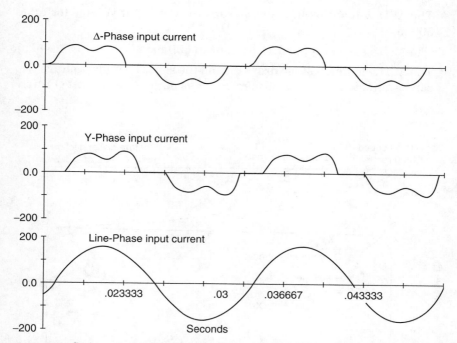

Figure 12.6 Series connected bridges 12-pulse converter for ASD. Secondary currents for wye and delta bridges. Transformer primary line current [12.6].
[© 1996, IEEE, reprinted with permission]

resultant rectifier input currents are shown in Figure 12.6 [12.6]. Each rectifier input current is typical for a six-pulse rectifier. The resultant 12-pulse rectifier line current has the 11th and 13th as the lowest-order harmonics (see Chapter 5).

Voltage disturbances

Three types of voltage disturbances affect ASDs: voltage sags, voltage interruptions, and voltage unbalance. In textile and paper mills, a brief voltage sag may potentially cause an ASD to introduce speed fluctuations that can damage the end product at severe cost. Furthermore, a brief voltage sag can also cause a momentary decrease in dc-link voltage, triggering an under-voltage trip or resulting in an over-current trip of the ASD [12.8].

The sensitivity of a 4-kW ASD, operating at rated speed and torque, to three-phase voltage sags is shown in Figure 12.7 [12.9]. The curves illustrate the following:

- The ASD will withstand voltage sags and interruptions to zero voltage for 10 to 20 ms.

- The ASD will withstand voltage sags to 70 percent voltage for up to 500 ms.

- The ASD's speed dropped 11 percent in 500 ms.

- The ASD's dc-link voltage drops during the sag. The under voltage trip can be adjusted as low as 50 percent, or as high as 90 percent of rated.

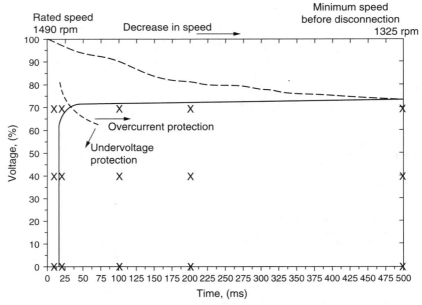

Figure 12.7 Voltage-time sensitivity of a 4-kW ASD to a balanced three-phase line-voltage sag—overcurrent and undervoltage protection [12.9].
[© 2005, IEEE, reprinted with permission]

- The overcurrent protection system can be triggered by increased current during the sag, or by high inrush current at voltage recovery to recharge the DC-link capacitor.

The effect of a constant-torque loading level on the sensitivity of the ASD in Figure 12.7 is shown in Figure 12.8 [12.9]. At 25-percent torque, the ASD can tolerate a voltage interruption of 50 ms and a voltage sag to 65 percent of the nominal line voltage up to 500 ms. Decreased loading results in better ride-through capabilities.

Voltage unbalance

Unbalance in the three-phase voltages supplying an ASD can produce the following effects [12.10]:

- The three-phase input diode rectifier reverts to single-phase mode, resulting in increased ripple in the dc-link voltage.

- The line current contains triplen harmonics.

- The diodes and the dc-bus capacitor are stressed by increased ripple, with possible life reduction.

- The motor develops uncharacteristic torque ripple, with increased noise and vibration.

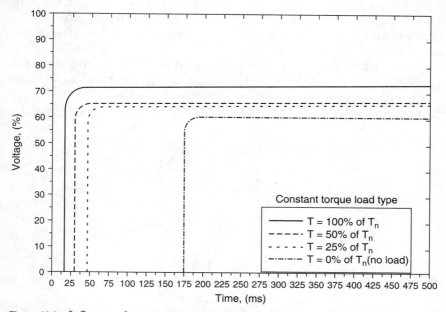

Figure 12.8 Influence of torque, 100 percent to 0 percent, on sensitivity of ASD to a balanced three-phase line-voltage sag [12.9].
[© 2005, IEEE, reprinted with permission]

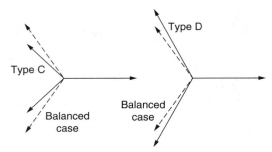

Figure 12.9 Definition phasor diagrams of Types C and D three-phase input voltage unbalanced [12.10]. [© 2004, IEEE, reprinted with permission]

One type of voltage unbalance termed Type C is shown in Figure 12.9 [12.10]. Phases *b* and *c* are shortened and phase shifted with respect to phase a. Unbalance is defined as:

$$\text{unbalance percent} = \frac{\text{Vavg} - \text{Vphase}}{\text{Vavg}} \times 100$$

$$\text{Vavg} = \frac{\text{Va} + \text{Vb} + \text{Vc}}{3}$$

Reference [12.10] describes an ASD subjected to a 2.5 -percent voltage unbalance, which reverted to the previously listed effects. The threshold for the condition depends on the line reactance and the dc-bus capacitor.

Figure 12.10 shows the three-phase input rectifier in the single-phase mode. The waveforms of the dc-bus voltage and the line current during the single-phase operation are shown in Figure 12.11. The input current consists of pulses each half cycle. The harmonic spectrum of the

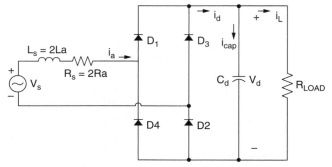

Figure 12.10 Effect of ASD input voltage unbalance. Equivalent circuit of rectifier stage during reversion to single-operation [12.10]. [© 2004, IEEE, reprinted with permission]

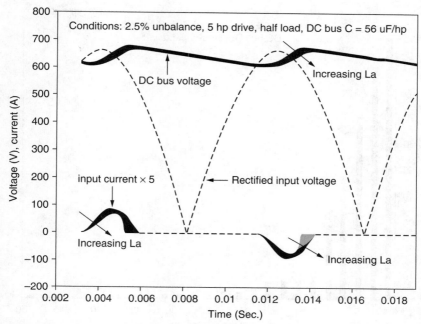

Figure 12.11 Operation under conditions of Figure 12.10. Waveforms of line current, DC bus voltage, and "raw-unfiltered" rectified AC line voltage for line reactance ranging, L_S 1 percent to 2 percent [12.10].
[© 2004, IEEE, reprinted with permission]

line current is shown in Figure 12.12. It includes the triplens as well as the 5th, 7th, 11th, and 13th harmonics, characteristic of three-phase operation [12.10].

Protective measures

Protective measures to maintain the operation of ASDs in the face of line voltage sags and interruptions include the following [12.7]:

- Restart
- Kinetic buffering
- Boost converter
- Active front end

The second, third, and fourth measures reduce the sensitivity of the ASD by increasing its tolerance to the depth and duration of line voltage sags and interruptions. External equipment, such as a UPS or E/G set, is required to maintain ASD operation for longer-duration voltage interruptions.

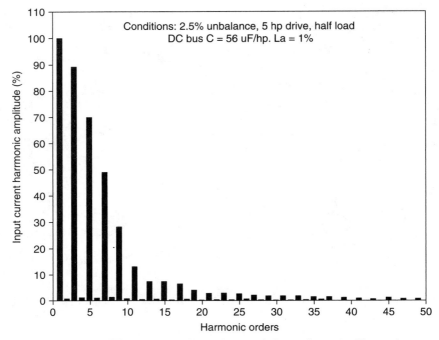

Figure 12.12 5 hp ASD. 2.5-percent line voltage unbalance, $L_S = 1\%$. Harmonic spectrum under unbalanced conditions [12.10].
[© 2004, IEEE, reprinted with permission]

Restart options are provided on commercial ASDs. Figure 12.13 shows a simplified version of the operation between power failure and recovery. The dc-bus voltage declines as the bus capacitor discharges while supplying the inverter. The inverter is shut down by the dc-bus under-voltage trip. The motor coasts down. The dc-bus voltage recovers when power is restored. The motor and inverter will only restart on a reset command [12.11].

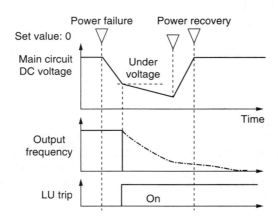

Figure 12.13 Commercial ASD operations following power failure, inverter shutdown by DC-bus undervoltage trip. No reset command. Motor coasts to a stop [12.11].

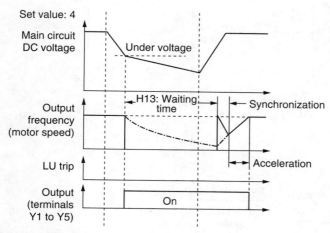

Figure 12.14 Commercial ASD operation following power failure. Power recovery, reset command, inverter restarted. Motor returned to former speed [12.11].

Figure 12.14 shows the operation of the ASD following a power failure and recovery when a reset command requires the inverter to restart and bring the motor back to its pre-failure speed. Finally, Figure 12.15 shows the operation with a high inertia load and kinetic buffering.

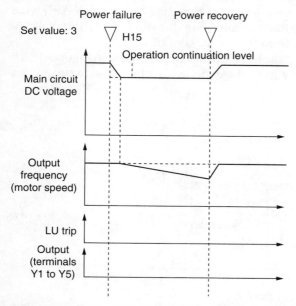

Figure 12.15 Commercial ASD operation following power failure. High inertia load. Kinetic buffering to maintain dc-bus voltage as motor speed declines. Motor returned to former speed upon power recovery [12.11].

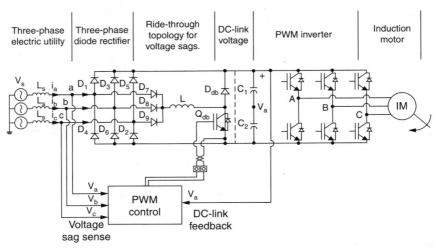

Figure 12.16   Integrated boost converter to maintain DC-bus voltage and ASD operation during voltage sag [12.3].
[© 2006, IEEE, reprinted with permission]

As the motor decelerates, the inverter transfers kinetic energy via the motor, acting as a generator from the mechanical load to the dc-bus capacitor. The method is suitable for a non-speed-sensitive load, such as a fan, but not a machine in a speed-sensitive process [12.11]. The measured voltage tolerance curve in Figure 12.17 for a 4-kW ASD with kinetic buffering shows that the tolerable duration of the line voltage interruption is extended from 0.02 to 0.80 s [12.7].

Under a line-voltage sag, the boost converters, as shown in Figures 12.16 and 12.18, transfer energy from the depressed supply line to the dc-bus

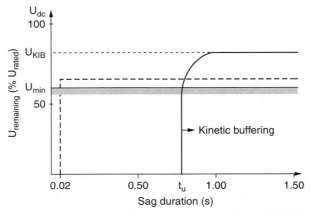

Figure 12.17   Measured voltage tolerance curve for a 4-kW ASD having a large inertia load, with and without kinetic buffering. Activation set at 90 percent rated dc-bus voltage [12.7].

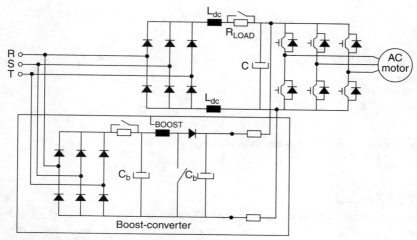

Figure 12.18 Parallel boost converter for DC bus of ASD in Figure 12.5 [12.7].

of the ASD. The tolerable sag duration depends on the power rating of the converter compared to the ASD. The curves in Figure 12.19 show tolerable sag duration extended from 0.02 to 2.0 s, and with a dip tolerance down to 25 percent of rated voltage. Obviously, the boost converter is not effective for a complete line voltage interruption unless a supplementary energy source is provided [12.7].

An ASD with an active front end is shown in Figure 12.20. The IGBTs in the rectifier portion are designed to extract power from the supply line for voltage sags down to 25 percent of rated voltage for durations of up to 30 s. The supply line must be capable of providing the current for the motor to deliver its required power during the sag [12.7].

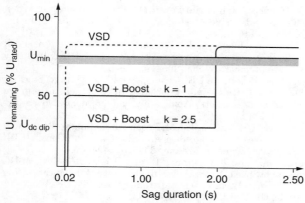

Figure 12.19 Voltage tolerance curves for the boost to the ASD in Figure 12.18. k = rating of boost converter/rating of ASD [12.7].

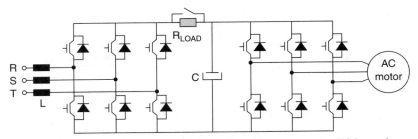

Figure 12.20   An ASD with an active front-end rectifier to support a DC-bus voltage during a line-voltage sag [12.7].

Summary

Induction motors alone (and in ASDs) represent a large part of the three-phase electrical load in commercial and industrial facilities. They are affected by line voltage sags and interruptions. Protective methods are available to maintain operation during voltage sag and to restart after voltage interruptions.

References

[12.1] A. E. Fitzgerald, C. Kingsley, and A. Kusko, *Electric Machinery*, 3rd edition, McGraw-Hill, New York, 1975.

[12.2] J. C. Gomez, M. M. Morcos, C. A. Reineri, and G. N. Campetelli, "Behavior of Induction Motor Due to Voltage Sags and Short Interruptions," IEEE Trans. on Power Delivery, vol. 17, no. 2, April 2002, pp. 434–440.

[12.3] M. Baran, J. Cavaroc, A. Kelley, S. Peel, and Z. Kellum, "Stresses on Induction Motors Due to Momentary Service Interruptions," 2006 IEEE Industrial and Commercial Power Systems Technical Conference, April 30–May 5, 2006, Detroit, MI.

[12.4] "NEMA Standard Publication No. MG 1-1998," Art 20.12, Number of Starts.

[12.5] A. Kusko and J. Y. Ayoub, "New Concepts in Large AC Motor Protection," EC&M, March 2006, pp. 22–24.

[12.6] J. W. Gray and F. J. Haydock, "Industrial Power Quality Consideration When Installing Adjustable Speed Drive Systems," IEEE Trans. on Industry Applications, vol. 32, no. 3, May/June 1996, pp. 95–101.

[12.7] K. Stockman, M. Didden, F. D. Hulster, and R. Belmans, "Bag the Sags, Embedded Solutions to Protect Textile Processes Against Voltage Sags," IEEE Industry Applications Magazine, September/October 2004, pp. 59–65.

[12.8] J. L. Duran-Gomez, P. N. Enjeti, and B. Okwoo, "Effect of Voltage Sags on Adjustable-Speed Drives: A Critical Evaluation and an Approach to Improve Performances," IEEE Trans on Industry Application, vol. 35, no. 6, November/December 1999, pp. 1440–1449.

[12.9] S. Z. Djokic, K. Stockman, J. V. Milanovic, J. J. M. Desmet, R. Belmans, "Sensitivity of AC Adjustable Speed Drives to Voltage Sags and Short Interruptions," IEEE Trans. on Power Delivery, vol. 20, no. 1, January 2005, pp. 494–505.

[12.10] K. Lee, G. Venkataramanan, and T. M. Jahns, "Design-Oriented Analysis of DC Bus Dynamics in Adjustable Speed Drives Under Input Voltage Unbalance and Sag Conditions," 2004 35th Annual IEEE Power Electronics Specialist Conference,

[12.11] AF-300G11 Users Guide, GE Fuji Drives USA, 2000.

Standby Power Systems

The term standby power systems describes the equipment interposed between the utility power source and the electrical load to improve the reliability of the electric power supply to the load. In previous chapters, we have described the deficiencies in the electric power supply, such as voltage sags and interruptions, and their effect on individual types of loads. The loads that require standby power can range from a single personal computer supplied by a battery-powered UPS to a large data processing center. In this chapter, we will describe the principal components used in standby power systems—namely, UPSs, transfer switches, and engine-generator (E/G) sets.

Principles: Standby Power System Design

The design of a standby power system for a specific electric-power load requires consideration of the following factors:

- **Reliability:** The questions are
  - What is the power quality of the utility source?
  - What is the cost of a power-quality event?
  - Will the system be required to operate continuously (24/7)?
  - What reliability of the electric power supply is required in terms of MTBF, minutes per year unavailability, Tiers, or other measures?
- **Maintenance:** The questions are
  - Can the load be shut down to perform maintenance or equipment replacement on the standby power system?
  - Can the standby power system be designed as online, or standby-redundant, for reliability or maintenance?

- **Expansion:** The questions are
  - Is expansion or modification of the load planned with or without shutdown?
  - Is expansion of the standby power system planned in terms of increased UPS-kVA rating, energy-storage capacity (batteries) or engine-generator kVA?
- **Personnel:** The questions are
  - What type of monitoring system is planned? Onsite, contract-type, alarm equipment?
  - What level of training will be given to personnel? Instruction regarding the complexity of the system? Will the training be done by vendors?
- **Cost:** The questions are
  - How much are measures worth for a standby power system over the basic unprotected supply of power to the load?

Components to Assemble Standby Power Systems

Components have been described in the previous chapters to correct for utility and internal-plant power deficiencies. These components can be assembled into standby power systems for critical loads and systems. They include the following [13.1]:

- **Transformers:**
  - Constant-voltage (CV) transformers for voltage regulation
  - Phase-shifting transformers for harmonic cancellation
  - Power distribution unit (PDU) transformers to supply individual loads
- **Dynamic Voltage Compensators:**
  - Compensation for short-time voltage sag and interruption
  - Extended time operation
  - Single-phase, three-phase loads
- **Uninterruptible Power Supplies (UPSs):**
  - Single-phase, battery support
  - Three-phase, battery support
  - Flywheel energy storage
  - Fuel-cell energy storage
- **Support Equipment:**
  - Engine-generator set
  - Gas-turbine generator set
  - Transfer switch, E/G set
  - Transfer switch, utility feeders

Sample Standby Power Systems

The design of practically every standby power system using the preceding list of components is unique. Each design is based upon the total electrical load (kVA), the reliability and number of utility feeders, the space available, the reliability requirements for the load power, the selection of the UPS modules, PDUs, transfer switches and E/G sets, the dollars available, and other factors. Sample basic systems in order of cost and complexity include the following:

1. **Dynamic Voltage Compensator plus a PC:** The simplest combination of a standby power system and a PC is shown in Figure 13.1. The compensator only protects the PC from short-time utility voltage interruptions up to three cycles (50 ms) and sags to 50 percent for up to 2 s.

2. **UPS plus a PC:** This configuration (as shown in Figure 13.2) of a single-phase battery-powered UPS supplying a single personal computer, is the most common arrangement for preventing voltage distortion, sags, surges, and interruptions from affecting the PC. Typical UPSs of 300-W rating, 5-min battery run time, and 120-V terminals in and out are commercially available. The UPSs are line-interactive (Delta) or double-conversion (online). The PC plus UPS can be used in an office, home, factory, or as part of an industrial control or monitoring system.

3. **UPS plus a server and PCs:** The battery-powered UPS can supply multiple PCs from the load terminals of a three-phase UPS, as shown in Figure 13.3, or a local area network (LAN), as shown in Figure 13.4. The UPS would be rated in the 2-to 10-kVA range, typically 208/120-V output, and 10-min battery run time. The UPS must have a sufficient power rating to provide the start-up inrush current to the inputs of the PCs, servers, and printers. The UPS will shut down for utility power interruptions exceeding the battery ampere-hour run time. For maintenance work on the UPS, such as battery replacement or repair, either the UPS is shut down, or the load is transferred to the utility source with a synchronized by-pass switch. Some UPSs utilize "hot swappable batteries."

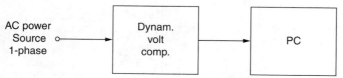

Figure 13.1 Block diagram of a dynamic voltage compensator supplying a PC.

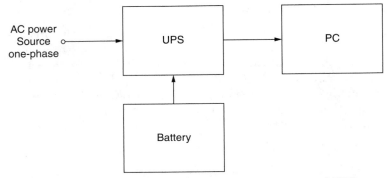

Figure 13.2 Block diagram of a single-phase battery-powered UPS sup-plying a PC.

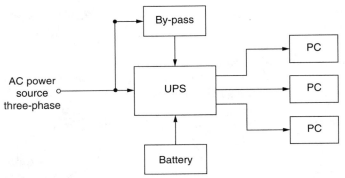

Figure 13.3 Block diagram of a three-phase battery-powered UPS supplying multiple PCs.

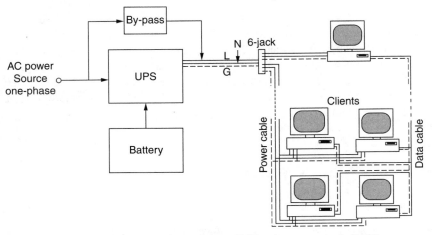

Figure 13.4 Battery-powered UPS supplying a PC local area network (LAN).

4. **UPS plus a Transfer Switch and an E/G Set:** An elementary standby power system to withstand long utility power interruptions and long UPS maintenance periods is shown in Figure 13.5. The system operates in the following ways:

- Normal-utility service. The UPS receives power from the UPS bus and delivers power to the load. The battery has been charged by utility power.

- Utility service is interrupted. The UPS receives power from its battery and continues to deliver power to the load. The E/G set starts. The transfer switch operates to supply generator power to the UPS. The UPS bus is transferred to the E/G set. The battery recharges. When utility service is restored, the process is reversed.

- UPS must be taken out of service. The E/G set is started. When it is up to speed, voltage, and frequency, the transfer switch operates. The UPS by-pass switch is closed to supply the load from the E/G set. The UPS is taken out of service. When the UPS is to be returned to service, the process is reversed.

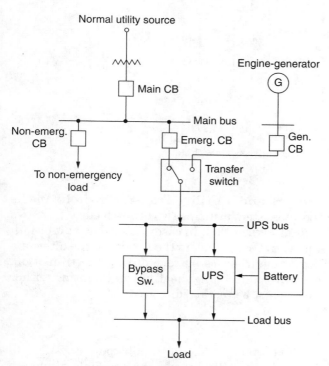

Figure 13.5 A battery-powered UPS system: utility source, engine-generator set, transfer switch, and by-pass switch.

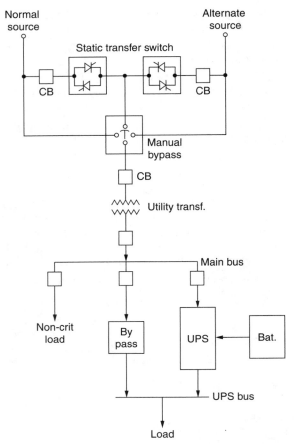

Figure 13.6 Two primary-voltage utility feeders with static transfer switch.

5. **UPS plus Two Feeders and a Utility Transfer Switch:** An elementary standby power system utilizing two utility feeders from two substations is shown in Figure 13.6. In case of a feeder fault or failure, the utility transfer switch transfers the service to the alternate feeder. The UPS operates from its battery during the transition. Other loads, such as air conditioning, heating, and lighting, either continue to operate or must be restarted.

Engine-Generator Sets

Engine-generator sets (E/G sets) are used in standby power systems to extend the operation time of the UPS beyond the available discharge time of the batteries. For large UPSs (for example, larger than 100 kVA), the battery time is usually only several minutes. The transfer from utility line to E/G set, including starting and stabilizing the generator for frequency and voltage, can be accomplished in about 10 s. Sometimes a short time delay is introduced in starting to avoid too frequent E/G starts for short-time voltage sags and interruptions that can be handled by the batteries of the UPS.

Engine-generator sets are available from 50 to 2500 kVA. A 218-kVA diesel-engine generator set is shown in Figure 13.7 [13.1]. For larger loads, it is common to employ multiple E/G sets operating in parallel to obtain reliability through redundancy and to have a set available for maintenance. In addition, electric power is required for non-UPS load—that is, air conditioning, lighting, and water pumps. Combustion turbines are also employed to drive generators in standby applications.

Standards

The major problem for the installation of E/G sets is environmental. The sets produce noise, exhaust, and vibration; they require coolant air or water, fuel, and fuel systems; they are heavy and require adequate foundations. A list of codes and standards governing E/G sets as an alternative to utility power includes the following:

- ANSI/IEEE 446-1995, "Recommended Practice for Emergency and Standby Power Systems for Industrial and Commercial Applications" [13.2]

Figure 13.7   Diesel engine-generator set. Rating: 175 kW, 0.8 PF, 480 Y / 277 V, 60Hz [13.1].

- NFPA 110-2002, "Standards for Emergency and Standby Power Systems" [13.3]
- NFPA 70-2005, " National Electrical Code" specifically [13.4]
  - Art. 700, "Emergency Systems"
  - Art. 701, "Legally Required Standby Power Systems"
  - Art. 702, "Optional Standby Power Systems"

Component parts of an E/G set installation

The component parts or subsystems of an E/G set are shown in Figure 13.8. Detailed information is given in *On-Site Power Generation: A Reference Book*, fourth edition [13.5]. The parts are described as follows:

- **Engine:** Gasoline, diesel, internal combustion engine, or a combustion gas turbine. Turbine requires minutes to start and acquire load, compared to seconds for an internal combustion engine.

- **Generator:** Three-phase salient-pole (1800 r/min or 3600 r/min) exciter for field current and damper windings for parallel operation.

- **Fuel:** Gasoline, diesel oil, natural gas, or another. Requires day tank and offsite tank, pumps, piping, vents, and filters.

- **Coolant:** Radiator and fan mounted on the engine, or external to the engine, with ducts to the engine.

- **Exhaust:** Muffler and exhaust piping to discharge gases and control noise.

- **Starter:** Electric motor, battery, and charger, or pneumatic means and high-pressure air storage. Start initiated by signal from manual switch or transfer switch.

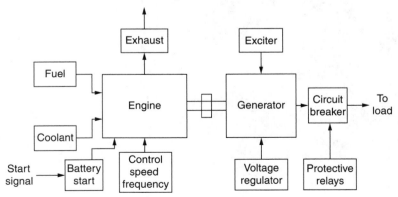

Figure 13.8   Block diagrams. Functions of E/G set.

- **Control:** Electronic isochronous governor for independent operation and when synchronized to other generators and/or to the utility line. Active power load division.

- **Generator Protection:** Relays to monitor load current, terminal voltage, reverse power, frequency, over temperature, to operate alarms and trip the circuit breaker.

- **Voltage Regulator:** Electronic regulator for the exciter or main field current to regulate terminal voltage. Reactive power load division.

- **Synchronizer:** Controls engine speed and circuit breaker closing when the E/G set is to operate in synchronism with the utility line and/or other E/G sets.

- **Circuit Breaker:** Electrically operated. Rated for generator overload current and for maximum three-phase short-circuit current.

Operation. The E/G set is a universal, useful, piece of equipment at a site whose output supplements utility service. Its operation includes the following functions:

- **Loads:** Include UPS, lighting, air conditioning, heating, fans, pumps, electronic equipment, and appliances. To a degree, loads can draw non-sinusoidal current (harmonics) and be unbalanced. The generator must have low subtransient reactance (damper windings) and/or be oversized for harmonic load current.

- **Emergency Operation:** The controls of the transfer switch order the E/G set to start when an interruption in utility voltage is detected (or a manual test signal is entered). When the generator voltage has stabilized at the correct amplitude and frequency, the transfer switch transfers the load to the E/G set. Starting and transfer usually takes about 10 s. When utility power is restored, the controls allow the E/G set to run for about 30 min before the transfer switch transfers the load back to the utility power. Multiple engine-generator sets start in sequence.

- **Testing:** E/G sets for emergency service should be tested about once a week, and allowed to run at least 30 min. The generator can be loaded with one of the following: (1) a dummy load, (2) a facility load, or (3) it can be synchronized to the utility line. The test can be initiated manually or automatically from a programmable controller.

- **Non-emergency Standby:** The E/G can be used to provide power to equipment or facilities when the normal power source is not available—for example, when they are under construction or maintenance work is being done. The UPS can be placed on by-pass. Adequate fuel must be available.

- **Parallel Operation:** Two or more E/G sets can be operated in parallel for redundancy, to secure more than one set's power, and to handle large motor starting. The governors and voltage regulators must insure parallel operation at the required voltage and frequency, as well as active and reactive power division.

Transfer switches

The purpose for a transfer switch is shown in Figure 13-5. The switch connects the load, in this case a UPS, to either the utility source or to an E/G set. The operation is usually conducted when power from the utility line is interrupted and the output of the E/G set must supplement the capacity of the batteries of the UPS. The transfer switch has other functions, such as transferring between two feeders or transformers, and transferring power to a motor from a failed feeder to an alternate feeder. The subject is covered extensively in reference [13.5].

Standards.   Pertinent standards on transfer switches include the following:

- NEMA, "AC Automatic Transfer Switches," ICS 2-44 [13.6]
- NFPA, "National Electrical Code," NFPA 70-2005 [13.4]
- UL, *Standard for Automatic Transfer Switches*, fourth edition, UL 1008 [13.7]
- NFPA, "Emergency and Standby Power Systems," NFPA 110-2005 [13.3]

Types of transfer switches.   Several types of transfer switches are available and include the following:

- **Manual:** The transfer switch can consist of a double-throw multi-pole switch, or two mechanically interlocked circuit breakers or contactors.
- **Automatic Electromechanical**
  - Three-pole or four-pole to switch three phases with or without neutral.
  - Open or closed transition, which usually requires utility permission when switching E/G sets with respect to a utility line.
  - Controls, including voltage sensing, engine starting and shutdowns.
- **By-pass:** Incorporates a by pass switch section that can connect the preferred line to the load while the transfer switch is removed for maintenance, as shown in Figure 13.9 [13.5].

Figure 13.9 Automatic transfer switch with optional bypass- isolation switch [13.5].

- **Solid State:** Utilizes thyristor or GTO AC switches in each leg, as shown in Figure 13-6. Transfer switches have been built for switching utility feeders rated up to 34 kV.

Applications.   Applications for transfer switches include nearly every area of electrical application, particularly those devoted to high reliability and safety. These include the following:

- **Utility to E/G Set:** A generic diagram is shown in Figure 13.10. The shape of the movable contact is typical to obtain a short travel distance and no ambiguity of contact positions.

- **UPS Supply:** Transfer from a utility line to an E/G set to supplement the UPS battery. Data centers employ up to 10,000 kVA of E/G sets and suitable transfer switches.

- **Electric-Motor Transfers:** For large induction motor–driven fans and pumps, as in power plants, transfer switches are employed where two alternate transformers or feeders are provided.

- **Lighting:** Alternate and emergency lighting systems supplied by transfer switches from batteries, inverters, and alternate feeders.

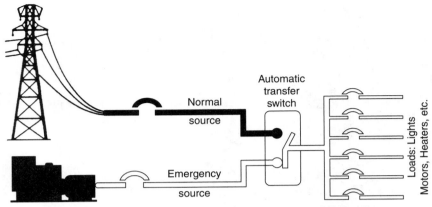

Figure 13.10   Normal/emergency sources. Automatic transfer switch [13.5].

Summary

Standby power equipment provides the means to improve the reliability of the utility electric power supply to critical load equipment. Alternate feeders and E/G sets utilizing transfer switches are the principal means of providing both power and standby power to UPSs, health care facilities, and telecommunications systems. The design requires consideration of all aspects of the existing utility supply and the desired objectives of the high-reliability system.

References

[13.1] A. Kusko, *Emergency/Standby Power Systems*, McGraw-Hill, New York, 1989.
[13.2] ANSI/IEEE 446-1995, "Recommended Practice for Emergency and Standby Power Systems for Industrial and Commercial Applications".
[13.3] NFPA 110-2005, "Standard for Emergency and Standby Power Systems".
[13.4] NFPA 10-2005, "National Electric Code".
[13.5] EGSA, *On-Site Power Generation Reference Book*, 4th edition, 2002.
[13.6] NEMA ICS 2-447, "AC Automatic Transfer Switches".
[13.7] UL1008, *Standard for Automatic Transfer Switches*, 4th edition.

14

Power Quality Measurements

In this chapter, we shall discuss practical issues related to making power-quality measurements. Any measurement job requires the proper tools to do that job. The monitoring of power quality at the supply end, at distribution and transmission, and at the load end has become of major interest in recent years. In the following, we will discuss various types of equipment used in power-quality measurements, and the proper use and limitations of such equipment.

Multimeters

Multimeters are used to measure voltages, currents, and resistances. Typical applications of multimeter measurements are line-voltage measurements and imbalance measurements. In making voltage and current measurements, generally we're interested in measuring the rms value of the waveforms. Some subtleties are associated with this measurement.

In one type of multimeter, the input waveform is rectified and averaged, as shown in Figure 14.1a. This method works well for pure sine waves (where the rms value is the peak value divided by the square root of two), but for waveforms with harmonics, this method is prone to errors. Error is especially high for waveforms with a high crest factor.

Several types of multimeters produce "true-rms" readings—that is, they accurately report the rms value of waveforms with harmonic distortion. One type of true-rms multimeter employs an rms-to-DC converter (Figure 14.1b). Several IC manufacturers produce rms-to-DC converter integrated circuits.[1] Another method of producing true-rms readings involves a thermal measurement method (Figure 14.1c) [14.1].

---

[1] See, for example, Analog Devices and Linear Technology.

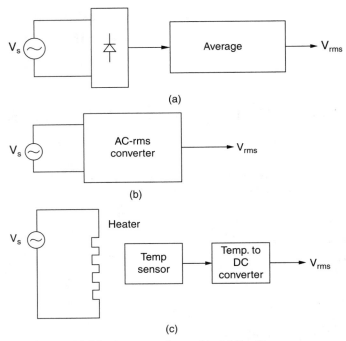

Figure 14.1 Multimeter types for making AC voltage measurements. (a) Rectify and average. (b) Analog computation. (c) Thermal.

True-rms multimeters are more accurate in the measurement of the rms value of periodic waveforms.

Oscilloscopes

Today's oscilloscopes have more than enough bandwidth to display the harmonics present in all power-quality events. For instance, the 20th harmonic of the line frequency in the U.S. is 1200 Hz, which is much lower than the bandwidth of even an inexpensive oscilloscope. Impulsive transients (like that caused by lightning strikes or power switching) typically last a few microseconds. It is well within the capability of oscilloscopes to measure and display these waveforms with good accuracy.

Scopes are often used in conjunction with current probes[2] to measure AC line currents. Often, differential voltage probe[3] modules are used to

---

[2] See, for example, the Tektronix P6042 current probe, which has a response from DC to 50MHz.

[3] The differential probe overcomes the practical problem of using a single-ended scope probe, which has a signal input and a ground clip.

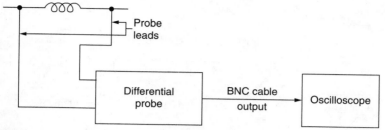

Figure 14.2 Measurement system for making high voltage differential measurements.

measure the voltage difference between two points in a system. Many manufacturers make high voltage differential probes that are compatible with oscilloscope inputs. Typical specifications for differential probes[4] are

- Bandwidth: 100 MHz
- Differential voltage: 5000 V
- Common-mode voltage: 2000 V

A typical system for making high voltage differential measurements is shown in Figure 14.2. We see that the output of the differential probe is a 50-Ω BNC cable[5] that interfaces with an oscilloscope. The differential probe has a conversion ratio (millivolts per volt) that brings down the signal levels to the oscilloscope to a manageable level.

Current Probes

Current probes come in several different types. Hall-effect current probes (like that used in the Tektronix P6042) detect all the way down to DC by utilizing the "Hall effect." Hall devices are semiconductors that generate an output voltage in response to a magnetic field created by a current. Hall probes have the advantage that they provide output at DC. Unfortunately, they are prone to drift and should be calibrated prior to use.

Other types of current probes use current transformers (CTs). CT current probes do not give an output at DC since they are based on transformer action. A typical test setup using a CT and an oscilloscope is shown in Figure 14.3. CT current probes give a conversion ratio (volts per amp) when terminated in the proper resistance—in most cases 50 Ω.

---

[4] See, for example, the Tektronix P5200 high-voltage differential probe.

[5] BNC cables are coaxial cables used in instrumentation and video. A BNC cable is characterised by its transmission line characteristic impedance, which is typically 50 or 75 Ω

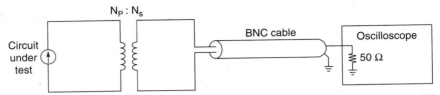

Figure 14.3 A measurement system for making high current measurements using a CT and an oscilloscope.

Search Coils

The magnetic field produced by a current can be detected by a "search coil" as shown in Figure 14.4. The voltage output of the search coil is found using

$$v_o = NA\frac{dB}{dt}$$

where N is the number of turns in the search coil, A is the area of the coil, and dB/dt is the time rate of change of the magnetic flux density perpendicular to the plane of the search coil. In the setup of Figure 14.4, the magnetic field H and the magnetic flux density B are found using

$$H = \frac{i}{2\pi r}$$

$$B = \mu_o H$$

where r is the mean radius from the current-carrying wire to the search coil, and μ_o is the magnetic permeability of free space. Note that this measurement can be affected by stray time-varying fields from sources other than the current-carrying wire being measured.

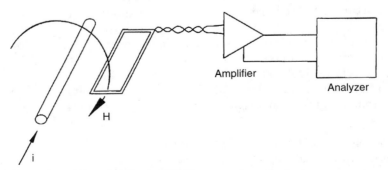

Figure 14.4 A search coil setup [14.2].
[© 1992, IEEE, reprinted with permission]

Power-Quality Meters and Analyzers

A number of manufacturers make "power-quality meters" and "power-quality analyzers," which are instruments similar to oscilloscopes,[6] but that have a number of functions particularly suited to making power-quality measurements. In measuring power quality, we must be able to trigger on events that are not continuous (such as a voltage sag) or transient (like that caused by lightning strikes or utility faults). For instance, typical power-quality meters/analyzers and analyzers have the following functionalities:

- Datalogging—capturing waveforms in real-time for later display
- The ability to trigger on power-quality events such as sags, swells, or transients
- Calculation of power-quality metrics such as total harmonic distortion in real-time
- Spectrum analysis
- Inputs for high-voltage probes and high-current probes

Numerous factors [14.3] should be considered when selecting a power-quality measurement meter/analyzer, including:

- The number of channels (for instance, single-phase or three-phase)
- The input voltage range
- The current measurement range
- Isolation
- Communication capabilities (For example, can the instrument be networked or tied to a stand-alone computer?)

The block diagram of a typical system with a power-quality analyzer is shown in Figure 14.5. The power-quality analyzer A/D converts system voltages and currents. Power indices such as THD, harmonic content, and the like are calculated in real-time. In addition, the analyzer has on-board memory so data may be saved for future analysis.

The output of a typical power-quality analyzer is shown in Figure 14.6, where we see the time waveform of the line current in a copy machine (top trace) and the harmonics (bottom trace).

Current Transformer Analysis in Detail

A typical setup for high-current sensing is shown in Figure 14.7. The primary side current we are trying to measure is I_p, and is supplied by a bus bar. The CT steps the current down with the ratio $1:N_s$. We resistively

---

[6] See, for example, Fluke and Dranetz BMI, among others.

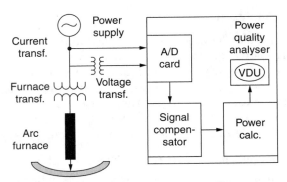

Figure 14.5 A block diagram of a power-quality analyzer [14.4].
[© 1997, IEEE, reprinted with permission]

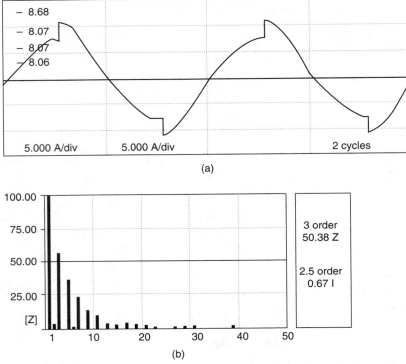

Figure 14.6 Power-quality analyzer measurements on a Xerox machine [14.5]. (a) Line current (b) Spectrum.
[© 2004, IEEE, reprinted with permission]

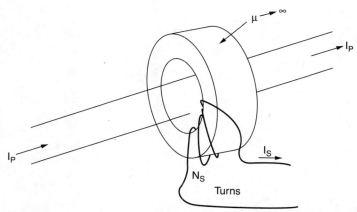

Figure 14.7 A typical current transformer for a high-current application. The primary of the transformer has one turn (due to the bus bar) and the secondary has Ns turns. The CT provides a current step-down.

load the secondary in order to convert the secondary current I_s to a voltage that we measure.

In order to sense the primary-side current, the CT secondary is loaded with a resistive load R_B (called the "burden"), as shown in Figure 14.8. This burden converts the secondary current I_S to a voltage V_{sense}, which is detected by electronics further downstream. The ideal input-output transfer function between the input current I_p and the sensed voltage V_{sense}, assuming no transformer parasitics, is

$$V_{\text{sense}} = \frac{I_p R_B}{N_s}$$

Of course, this transfer function assumes the transformer is ideal. A more realistic model of a current transformer is shown in Figure 14.9a. Elements of this model include

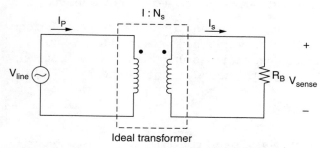

Figure 14.8 The ideal CT loaded with the "burden" resistor R_B.

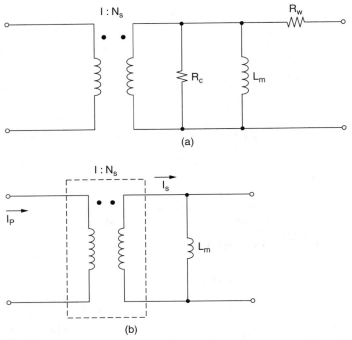

Figure 14.9 Models of a current transformer. (a) Includes parasitic elements such as core loss resistance R_c, magnetizing inductance L_m (referred to the secondary), and secondary winding resistance R_w. (b) Model resulting from the assumption that the core loss and secondary winding resistance have negligible effects.

- R_c: The core loss resistance, which models power dissipation due to eddy currents and hysteresis in the core
- L_m: The magnetizing inductance. This is the inductance you would measure if you connected up an impedance analyzer to the secondary side of the transformer, with the primary open-circuited.
- R_w: The copper winding resistance of the secondary winding.

In some cases, the core loss and secondary winding loss are negligible, and we can model the transformer as shown in Figure 14.9b.

A model[7] of the secondary circuit is shown in Figure 14.10. In this circuit, we've referred the primary current to the secondary side of the

---

[7] Of course, this more realistic view still ignores many important parasitic effects, including secondary winding resistance, core loss, and primary to secondary leakage inductance.

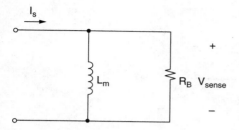

Figure 14.10 A secondary circuit. The current I_s is the primary current referred to the secondary through the turns ratio N_s.

transformer, and termed it I_s. The value of I_s is stepped-down from the primary and is

$$I_p' = \frac{I_p}{N_s}$$

We see that the sensed voltage V_{sense} depends not only on the value of the burden resistor R_b, but also on the value of the "magnetizing inductance"[8] L_m as measured on the secondary side. The transfer function relating the reflected primary current I_p' to the sensed voltage V_{sense} in the Laplace domain is

$$\frac{V_{\text{sense}}(s)}{I_p'(s)} = \frac{L_m s}{\dfrac{L_m}{R_B} s + 1}$$

The magnitude and phase[9] of this transfer function are

$$\left| \frac{V_{\text{sense}}(\omega^2)}{I_p'(\omega^2)} \right| = \frac{L_m \omega}{\sqrt{\left(\dfrac{L_m}{R_B}\omega\right)^2 + 1}}$$

$$\angle \frac{V_{\text{sense}}(\omega)}{I_p'(\omega)} = \frac{\pi}{2} - tan^{-1}\left(\frac{\omega L_m}{R_B}\right)$$

Note that at very low frequencies (well below R_B/L_m), the transfer function magnitude approaches zero. This is as expected, because a transformer does not pass currents at DC. At very high frequencies, the magnetizing inductance L_m essentially becomes an open circuit and all the current passes through the R_B burden resistor.

---

[8] The magnetizing inductance is the inductance you would measure if you put the secondary winding on an impedance analyzer.

[9] Note that the phase is expressed in radians. The conversion to degrees is degrees = radians $\times$ (360/2π).

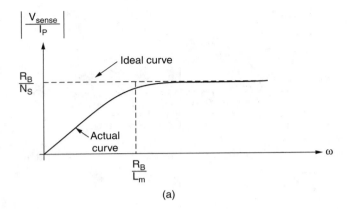

(a)

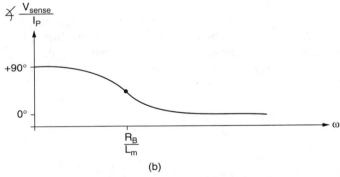

(b)

Figure 14.11 The transfer function relating sensed voltage V_{sense} to input primary current I_p. Note that the ideal input-output transfer function has a flat gain over all frequencies, and zero phase shift.

The magnitude and angle of this transfer function are shown in the Bode plots of Figure 14.11. Note that the transfer function of the current transformer approaches the ideal only for frequencies where $\omega \gg R_B/L_m$. Also note that there can be significant phase shift through the CT even if the magnitude of the transfer function is very close to the ideal. Let's examine the transfer function gain and angle in the case of $\omega \gg R_B/L_m$.

$$\left| \frac{V_{\text{sense}}(\omega)}{I_p'(s)} \right|_{\omega \gg \frac{R_B}{L_m}} \approx R_B$$

$$\angle \frac{V_{\text{sense}}(\omega)}{I_p'(\omega)} = \frac{\pi}{2} - \tan^{-1}\left(\frac{\omega L_m}{R_B} \right) \approx \frac{R_B}{\omega L_m}$$

Figure 14.12 Current probe measurements on a adjustable-speed drive.

Note that the angle is expressed in radians. This result shows that there can be significant phase shift through the transformer even if the magnitude is very close to the ideal.[10]

Figure 14.12 shows three current probes for making real-time measurements on a 600-V adjustable speed drive, used by one of the authors in a factory measurement in 2005. Instrument transformers are discussed in much detail in references [14.6] and [14.7].

Example 14.1: Current transformer. A 60 Hz current transformer has a primary current of 500 A and a turns ratio of 1000:1. The CT has a burden resistor of 1 ohm and a magnetizing inductance of 100 milli-henries. We'll find the 60 Hz gain and phase error for a current transformer with magnetizing inductance $L_m = 1$ millihenry and burden resistor $R_B = 1$ Ω. We'll also find the ideal output voltage level V_{sense}, assuming a transformer turns ratio of 1000:1, and the power dissipated in the burden resistor.

The ideal output voltage level is

$$V_{sense} = \frac{I_p R_{sense}}{N_s} = \frac{(500)(1)}{1000} = 0.5\,\text{V}$$

---

[10] This result is characteristic of first-order systems.

The magnitude of the sensed voltage is

$$|V_{\text{sense}}| = \left(\frac{I_p R_B}{N_s}\right)\frac{1}{\sqrt{\left(\dfrac{L_m \omega}{R_B}\right)^2 + 1}} = (0.5\,\text{V})(0.999648) = 0.4998\,\text{V}$$

Therefore, the magnitude error is only 0.04 percent. We find the angle[11] as follows:

$$\angle V_{\text{sense}} = 90° - \tan^{-1}\left(\frac{\omega L_m}{R_B}\right) = 90° - 88.5° = +1.5°$$

Therefore, the sensed voltage has a $+1.5°$ phase shift in relation to the primary current. The power dissipated in the burden resistor is due to the secondary current of 0.5A, and is

$$P_{\text{diss}} = I_s^2 R_B = 0.25\,\text{W}.$$

Example 14.2: CT error.   A CT is utilized in a data acquisition system and is used in the calculation of real power, reactive power, and the power factor in a single-phase system with sinusoidal voltages and currents. CT has an ideal magnitude response, but a positive phase shift of 5 degrees. Using data from this CT and other instrumentation, the data acquisition system calculates values of:

- Apparent power $S = 10\,\text{kVA}$
- Real power $P = 9\,\text{kW}$
- Reactive power $Q = 4.36\,\text{kVAr}$
- Power factor PF $= 0.9$ lagging

We shall determine the error in the four measurements due to the phase shift through the CT.

The calculation of apparent power is correct, since calculation of S does not depend on the phasing of the measured voltage and current. So, we find

$$S = 10\,\text{kVA}$$

---

[11] The $+90°$ is due to the inductor at low frequencies. For a purely inductive circuit, the voltage leads the current by 90 degrees.

Next, we'll look at the power factor. The data acquisition system cal-culated a power factor of 0.9, which means the angle between voltage and current is

$$\theta = \cos^{-1}(0.9) = -25.8°$$

However, we know that the actual angle is 5 degrees more (due to the error in phase shift from the CT). Therefore, the actual power factor is

$$PF = \cos(-30.8) = 0.86$$

We now find the real power:

$$P = (PF)(S) = (0.86)(10 \text{ kVA}) = 8.6 \text{ kW}$$

Next, we find the reactive power:

$$Q = \sqrt{S^2 - P^2} = 5.1 \text{ kVAr}$$

The result is that a relatively small phase shift can result in large cal-culation errors.

Summary

In this chapter, we have discussed some of the subtleties of power-qual-ity measurement, including selection of the proper measurement equip-ment. A successful power-quality measurement involves choosing the proper measurement tool, and using it correctly.

References

[14.1] J. Williams and T. Owen, "Performance Verification of Low Noise, Low Dropout Regulators," Linear Technology Application Note 83, March 2000. Available on the Web at http://www.linear.com/pc/downloadDocument.do?navId=H0,C1,C1003, C1040,D4172.

[14.2] IEEE, "IEEE Recommended Practices and Requirements for Harmonic Control in Electrical Power Systems," IEEE Std. 519-1992, revision of IEEE Std. 519–1981.

[14.3] Dranetz-BMI, *The Dranetz-BMI Field Handbook for Power Quality Analysis*, 1998

[14.4] B. Boulet, L. Karar, and J. Wikston, "Real-Time Compensation of Instrument Transformer Dynamics Using Frequency-Domain Interpolation Techniques," *IEEE Instrument and Measurement Technology Conference*, May 19–21, 1997, pp. 285–290.

[14.5] V. Kumar, P. S. Kannan, T. D. Sudhakar, and B. A. Kumar, "Harmonics and Interharmonics in the Distribution System of an Educational Institution—A Case Study," *2004 International Conference on Power System Technology (POWERCON 2004)*, November 21 and 24, 2004, pp. 150–154.

[14.6] IEEE, "IEEE Standard Conformance Test Procedures for Instrument Transformers," IEEE Standard C57.13.2-1991, published by the Institute of Electrical and Electronics Engineers.

[14.7] ____, "IEEE Standard Requirements for Instrument Transformers," IEEE Std. C57-13-1993, published by the Institute of Electrical and Electronics Engineers.

Index